Three

EXTRATERRESTRIAL DUST
Laboratory Studies of Interplanetary Dust

ASTROPHYSICS AND SPACE SCIENCE LIBRARY

VOLUME 181

EXTRATERRESTRIAL DUST

Laboratory Studies of Interplanetary Dust

Kazuo YAMAKOSHI

Institute for Cosmic Ray Research, University of Tokyo, Tokyo, Japan

Terra Scientific Publishing Company, Tokyo

Kluwer Academic Publishers / Dordrecht, Boston, London

Library of Congress Cataloging-in-Publication Data

Yamakoshi, Kazuo, 1936-
Extraterrestrial dust : laboratory studies of interplanetary dust / by Yamakoshi, Kazuo.
p. cm. -- (Astrophysics and space science library ; v. 181)
ISBN 0-7923-2294-0 (alk. paper)
1. Cosmic dust. 2. Marine sediments--Analysis. I. Title. II. Series.
QB791.Y36 1994
523.1'125--dc20 94-29414

Published by Terra Scientific Publishing Company (TERRAPUB),
302, 303 Jiyugaoka Komatsu Building, 24-17 Midorigaoka 2-chome,
Meguro-ku, Tokyo 152, Japan,
in co-publication with Kluwer Academic Publishers, Dordrecht, The Netherlands

Sold and distributed in the U.S.A. and Canada
by Kluwer Academic Publishers,
101 Philip Drive, Assinippi Park, Norwell, MA 02061, U.S.A.
in Japan by Terra Scientific Publishing Company (TERRAPUB),
302, 303 Jiyugaoka Komatsu Building, 24-17 Midorigaoka 2-chome,
Meguro-ku, Tokyo 152, Japan

In all other countries, sold and distributed
by Kluwer Academic Publishers,
P.O. Box 322, 3300 AH Dordrecht, The Netherlands

(This book is published by Grant-in-Aid for publication of Scientific Research Result of the Ministry of Education, Science and Culture of Japan)

Printed in Japan

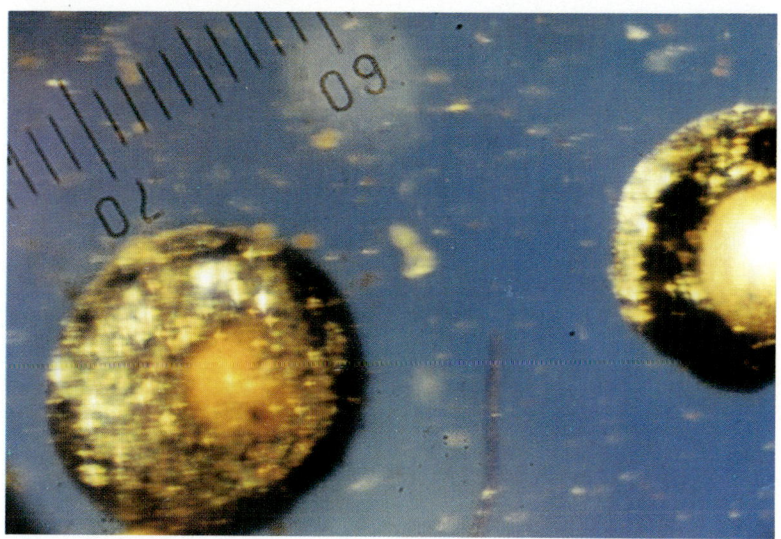

Photo 1. Black, magnetic iron spherules from deep sea sediments. Magnetite crust was broken so that the inner core can be seen, whose size is approximately 100 microns.

Photo 2. Magnetic, stony spherules collected from deep sea sediments, sometimes called "chondritic spherules". The size of the biggest one is 440 microns.

Kazuo Yamakoshi

Kazuo Yamakoshi is a professor at the Institute for Cosmic Ray Research, University of Tokyo, Tanashi, Tokyo, Japan. He received a B.S. in 1961, an M.S. in 1963 and a Ph.D. in Physics in 1966 from Gakushuin University. From 1961 to 1966 he studied the long-term variation of cosmic ray intensity using ^{14}C measurements of tree rings at the Institute for Nuclear Study, University of Tokyo. After receiving his Ph.D., he joined the cosmic dust research group at Kyoto University as a research associate. In 1972 he returned to the Institute for Nuclear Study, University of Tokyo, as an associate professor and worked as the head of the Cosmic Matter Division. In 1976 he moved to the Institute for Cosmic Ray Research, University of Tokyo. He has researched various features of cosmic spherules and has succeeded in estimating the cosmic ray exposure ages of cosmic spherules and in deducing their origins. In addition, he has also studied extremely low background radiation measurements. He has written two scientific monographs in Japanese: "Low Level Radiation Counting Techniques" (1979) and "Interplanetary Dust" (1984). He now is studying isotopic anomalies in primordial samples in carbonaceous chondrites, cosmic events within 10^7 yr using long-lived, cosmogenic radioisotopes and in-situ measurements of interplanetary dust and simulation laboratory measurements. For "Studies on Cosmic Matter in Deep Sea Sediments" he was awarded the Tanakadate Prize in 1986 from SGEPSS (Society of Geomagnetism and Earth, Planetary and Space Sciences, Japan). Also a golden medal of merit was given him from the Institute of Geochemistry, Chinese Academy of Science, in 1989.

PREFACE

A lucky discovery of black magnetic spherules from deep sea sediments in 1968 prompted me to study them. Since this discovery, the beautiful appearances of deep sea spherules have fascinated me and led to the start of scientific studies on their origin, composition and so on. Just twenty years after this discovery, the spherules have stimulated me to write this book. Despite the passage of such along time, many problems regarding their origins remain unsolved.

Until 1968, I, as a postgraduate student, was engaged in studying the long-term variations of cosmic ray intensity by measuring the ^{14}C (half-life = 5730 y) in individual tree rings of a large Japanese cedar tree.

^{14}C is produced by high energy cosmic rays in the Earth's atmosphere. After the nuclide is fixed in the biosphere through photosynthesis, it decays monotonously according to the exponential law. The effectiveness of the radio-carbon age-determination method was first verified by R. F. Libby.

The opposite logic was applied in our study to check the constancy of cosmic ray intensity in the past. The formula of radioactive decay in a closed system is given as follows;

$$N_t = N_o \cdot exp(-\lambda t),$$

where N_t is the atom number at time t, N_o is the initial atom number, and λ is the decay constant. Here N_t is an observable value, and λ is an universal constant. If we assume N_o is nearly constant in the past, we can obtain the age of a closed system, such as archaeological samples.

However, if t is given (we used tree rings) and we measure the N_t values very precisely, we obtain the N_o values at the respective "tree ring time". The N_o values are dependent on the production rates of ^{14}C by high energy cosmic rays in the atmosphere. Many physical, chemical and climatic factors (such as intensity variations of the magnetic fields of the Sun and Earth, changes of the exchange rates of carbon dioxide between atmosphere and ocean and/or changes of Galactic cosmic ray intensity) could change the N_o values over long historical periods. Therefore, we must conclude that the N_o values might be a time-dependent parameter: $N_o = N_o(t)$.

Precise determinations of ^{14}C in various tree rings revealed long-term variations of ^{14}C production rates, and the patterns of the variations promised faintly to show the causes.

"As the next step," Prof. H. Hasegawa of Kyoto University said in 1966, "studies on radioactive ^{59}Ni should be fruitful!"

The remarkably high content of nickel in deep sea sediments interested us. At that time, the major part of nickel in pelagic clays was supposed to be of extraterrestrial origin. ^{59}Ni (half-life = 7.6×10^4 y) decays only through the electron capture process, whose KX-ray energy is as low as 6.9 KeV! And it is a reaction product of solar alpha particles predominantly with ^{56}Fe;

$$^{56}\mathrm{Fe}(\alpha, \mathrm{n})^{59}\,\mathrm{Ni}.$$

The precise determination of ^{59}Ni at various depths in core samples, whose ages were already known from the paleomagnetic dating method, promised to give us the solar activities of alpha particle emitting phenomena (such as solar flares) in the past. However, because of low energy X-rays from ^{59}Ni, it was very difficult to detect them. Chemical extraction of the nickel fraction from sea bottom sediments was also very hard work for us. The leaching process of the dried sediments with large amounts of hydrochloric acid was carried out in many, big, plastic buckets, which we cynically called "bucket chemistry".

It was during a cruise on a research vessel of the University of Tokyo in 1968 to take dredged sediments and red clay core samples that, following the suggestion of a marine biologist, I was able to collect several brilliant black spherules among the gathered samples from a plankton net under a microscope. Thereafter, I collected a large number of spherules from magnetic fractions gathered from dredged deep sea sediments.

These spherules (except the glassy ones) had already been discovered by J. Murray during the Challenger Expedition carried out from 1874 to 1876 and reported in the famous "Challenger Report" with beautiful colored lithographs. J. Murray called them "cosmic spherules". Much work has been done on them during the century since their discovery.

However, nobody succeeded in confirming them clearly as being of extraterrestrial origin up to the 1970s. However, thereafter we have been inundated by works on the spherules: specific gravity, size distribution, chemical composition, isotopic anomalies, cosmogenic radionuclide detection and thermal degeneration processes during meteor flashing.

During these last two decades, spectacular events in space, such as the Apollo and Luna missions, and some epoch-making problems, such as the isotopic anomalies found in abnormal phases of Ca-Al-rich inclusions in carbonaceous chondrites, have interested strongly us.

In the 1970s, the Johnson Space Center of NASA began a laboured "mail order" of Brownlee's particles. The "Cosmic Dust Catalog" has been published periodically, and in the catalog we can find pictures, some parameters (such as size, colour, forms, transparency, etc.) and also XMA diagrams. In the XMA diagrams, the major elements of which the meteorites consist, such as Ca, Al, Mg, Fe, Ni, S and/or Cr, are detected in these microfine aggregate samples, which are classified in the catalogs as "code C", of cosmic origin.

We have also performed dust collection experiments with a large volume balloon in 1984. The flight duration was about 11 hours at an altitude of about 22 km. The processed air mass did not exceed 6×10^6 m^3. However, the flight time was so immediately after the El Chichon volcanic eruption that almost all the "dust" samples might be derived from volcanic ash.

From the 1950s dust counting was started by on-board detectors in space. After the failure of the piezo-electric microphone sensors, pressurized pad systems, mesh sensors for incoming dust directions and time-or-flight analyzers and so on have been successfully developed by scientists mainly from the USA and Germany.

In 1986, the Giotto and Vega missions obtained excellent data on the dust of Halley's Comet. These fruitful data were gained on the basic knowledge from fundamental experiments carried out by laboratory simulation experiments.

A series of interesting information on dust in interplanetary space has continued to give a strong impetus to researchers. After the Halley's Comet fever, we could begin to set up some projects in Japan for direct analyses and catch and return programs for interplanetary and cometary dusts.

In interplanetary space and around the Earth, we can recognize cosmic dust in various phases and phenomena: zodiacal lights, counterglow, meteor-flashing and cosmic spherules found in polar ice or pelagic sediments. In the following chapters, I will summarize the facts and results obtained using various observational and analyzing techniques and then compare them with one another.

For example, the annual rate of cosmic meteoroid accretion on the Earth and its size distribution can be evaluated by meteor observations with optical and radar techniques, spherule counting in rainwater, ice layers and/ or deep sea sediments, Ni or Ir content determinations of deep sea sediments and also radioactive measurements of cosmogenic radionuclides. Noble gas studies using deep sea sediments give us fruitful information about solar activities and/or the accretion rate of cosmic dust.

Up to now, many problems concerning the features of the interplanetary dust remain unsolved. This article will discuss comparative studies of the data obtained in various research fields and search for unified conclusions.

K. Yamakoshi

ACKNOWLEDGEMENTS FOR PERMISSION TO REPRODUCE THE DIAGRAMS USED IN THIS ARTICLE:

The author gratefully acknowledges permissions given by the authors and publishers (organizations) to reproduce the diagrams in this book which are included in the following articles; J. P. Bradley *et al.*, Nature 301(1983)473, J. P. Bradley, Science 226(1984)1432, D. E. Brownlee *et al.*, Science 191(1976)1270, D. E. Brownlee *et al.*, Nature 309(1984)693, H. Fechtig *et al.*, Ann. New York Acad. Sci. 119(1964)243, H. Fechtig *et al.*, "Cosmic Dust" (J. Willey & Sons, N.Y., 1978) edt'd by J. A. M. McDonnell, A. Fujiwara *et al.*, Nature 272(1978)602, R. Ganapathy *et al.*, Science 220(1983)1158, B. Hudson, Science 211(1981)383, D. Hughes, Space Research (1975) 333, S. Furukawa, MS thesis of Musashi-Industrial-Institute (1990), H. Iglseder *et al.*, The diagram of Munich Dust Counter (1990). T. Inoue *et al.*, EPSL 45 (1979) 181, O. A. Kirova, Ann. New York Acad. Sci. 119(1964)235, E. L. Krinov, Ann. New York Acad. Sci. 119(1964)224, R. M. MacQueen, Astrophys. J. 154(1968)1059, T. Maihara *et al.*, Proc. 85th IAU Colloquium (Marseille, 1985) pp. 56, M. Maurette *et al.*, Nature 328(1987)699, J. A. M. McDonnell *et al.*, "Cosmic Dust" (J. Wiley & Sons, N.Y., 1978) edt'd by J. A. M. McDonnell, J. A. M. McDonnell, Nature 309(1984)273, M. T. Murrell *et al.*, Geochim. Cosmochim. Acta 44(1980)2067, K. Nogami *et al.*, Geochem. J. 19(1985)101, K. Notsu *et al.*, EPSL 42(1978)903, D. W. Parkin *et al.*, Geophys. J. Roy. Soc. 1 (1966), T. Sasaki, Osaka Chigaku Kyoiku 5 (1983) 9, M. Takayanagi *et al.*, J. Geophys. Res. B12 (1987)1253, T. Yamamoto *et al.*, Publ. Astronom. Soc. Japan 31(1979)585, S. Yanagita *et al.*, Nucl. Phys. A303(1978)254, S. Yanagita *et al.*, EPSL 52(1981)259, M. E. Zolensky, [CDPET]; "Cosmic Dust Catalog" from NASA (1978).

ACKNOWLEDGEMENTS

The author is indebted to Mr. Keiji Oshida, the representative of Terra Scientific Publishing Company for his encouragement and management.

This book is supported financially by a Grant-in-Aid for Publication of Scientific Research Results from the Ministry of Education, Science and Culture of Japan in 1991.

I thank many authors, publishers and the representatives of many institutes and organizations for permission to reproduce their diagrams in this work. (See the acknowledgements in detail at the end of the preface.)

I am deeply grateful to my wife, Yukie Yamakoshi, for improvement of the English and scientific expressions in this book and for continuous encouragement.

CONTENTS

Chapter 1

Spatial Distribution of Interplanetary Dust

1.1 *Zodiacal light and counter glow*

Zodiacal light is a luminous zone like a tongue along the ecliptic orbit in the sky. It can be seen in the west after sunset and/or in the east before sunrise.

It is considered to be sunlight scattered by fine dust grains which are in gravitational orbits in the inner region of the solar system. Also, a vague elliptical patch of light in the clear night sky appears at the opposite site to the sun. It is called "counter-glow" (or Gegenschein in German). It is considered to be sunlight reflected by interplanetary dust clouds on the ecliptic plane spread behind Earth's orbit.

However, the detailed properties of the dust grains, such as size distribution, spatial concentration, shape, chemical composition, radiation reflection properties, are still being studied by astronomical observation.

Up to now, some models for these dust grains have been proposed: an earlier model, which was conceived by van de Hulst (1949) and improved by Oepik (1951), suggested a grain size distribution

$$n(r)dr = c \cdot r^{-p},$$

where the grain size range was 1~350 μm. In this size region, p was postulated as 2.6 to 3.0.

The polarization which is obtained in zodiacal light was interpreted as scattering from interplanetary electrons (Thomson's scattering). Blackwell and Ingham (1961) gave the figure of a few electrons per cc in the interplanetary space; however, the reflection power by an electron cloud is so small that the expected electron density in space is much smaller than that observed experimentally.

Giese (1962) and Weinberg (1962) proposed that the scattered intensity and polarization are caused by microfine dust grains near the solar region ("Mie" scattering) and estimated $p = 4.0$ and extended the size region to 0.1 μm. The Mie scattering process occurrs when the wavelength of the radiation is nearly equal to the size of the grains. The spatial distribution of zodiacal light suggests that the dust grains have modest inclinations and that their velocity relative to the Earth tends to be quite low, such as 2~5 km/sec.

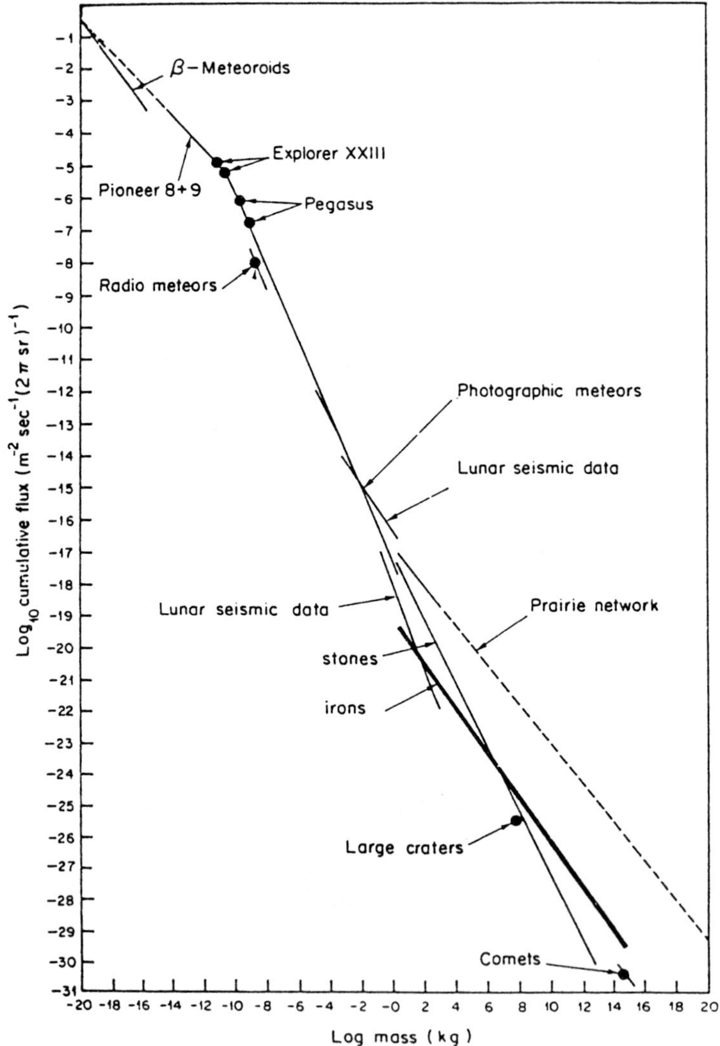

Fig. 1.1. Cumulative accretion rate of meteoroids onto the Earth's surface obtained data by various methods; per (m^2, 2π ster. sec.) (Dohnanyi 1978).

1.2 Accretion Rate of Cosmic Matter onto the Earth

The accretion rate of cosmic matter onto the Earth has been studied by various methods. Here one diagram is shown in Fig. 1.1, compiled by Dohnanyi (1978), in which he proposed a well fitting empirical formula for the mass range from 10^{-20} to 10^{+20} kg as a pre-atmospheric mass.

$$\frac{dN(M)}{dM} = 3.0 \times 10^{-18} \cdot M^{-13/6}, \quad \left(10^{-10}\, kg \le M \le 1 kg\right)$$
$$= 1.4 \times 10^{-11} \cdot M^{-1.5}, \quad \left(M \le 10^{-10}\, kg\right),$$

where $dN(M)$ is the influx per (m^2, sec., 2π ster.) into the Earth's atmosphere of meteoroids, which have a pre-atmospheric mass in the mass range of M to $M + dM$ kg.

The spatial distribution of the dust grains, $n(r)$, is expressed by Link *et al.* (1976) from the data obtained by the Helios A mission:

$$n(r) \propto r^{-1.3}, \quad 0.08\, AU \le r \le 1.5 AU$$

where $n(r)$ is the grain number density at a distance r from the Sun. Hanner *et al.* (1976) proposed a model which also fits the zodiacal light distribution in the interplanetary space:

$$n(r) \propto r^{-1}.$$

Recent data obtained with various observations and direct measurements in deep space show that the grain number density is nearly constant at between 2~16 AU (Greenberg, 1990).

REFERENCES

Blackwell D. E. and Ingham M. F. 1961 Month. Notice Roy. Astronom Soc. 122 113.

Dohnanyi J. S. 1978 Chap. 8, Cosmic Dust ed. J. A. M. McDonnell, (John Wiley & Sons)

Hanner M. S., Sparrow J. G., Weinberg J. L. and Beeson D. E. 1976 Interplanetary Dust and Zodiacal Light, Lecture Note vol. 48, ed. H. Elsaesser and H. Fechtig (Heidelberg: Springer) pp. 29.

Giese R. H. 1962 Space Sci. Rev. 1 589.

Greenberg J. M. 1990 Concluding Remarks in #126 IAU Colloquium in Kyoto.

Link H., Leinert C., Pitz E. and Salm N. 1976 Interplanetary Dust and Zodiacal Light, Lecture Note in Physics 48 ed. H. Elsaesser and H. Fechtig (Heidelberg: Springer) pp. 48.

Oepik E. J. 1951 Proc. Roy. Soc. Irish Acad Sci. 54A 165.

van de Hulst H. C. 1947 Astrophys., J. 105 471.
Weinberg J. L. 1976 Interplanetary Dust and Zodiacal Light, ed. H. Elsaesser and H. Fechtig (Heidelberg: Springer) Lecture Notes 48 pp. 3.

Chapter 2

Motion and Origins of Interplanetary Dust

2.1 Poynting-Robertson Effect

The dust moving in Kepler's orbit in interplanetary space is affected by solar radiation pressure and subsequently goes "down" spirally into the Sun. In general, we say that this spiral motion of the interplanetary dust is caused by the "Poynting-Robertson Effect" (Robertson, 1937 and Wyatt and Whipple, 1950).

In a coordinate system, in which the dust is at rest, the solar radiation appears not to irradiate the dust from the center of the Sun, but instead to come from such a direction as to oppose the transverse component of the dust motion. After absorption of the sunlight, the temperature of the dust increases, and the absorbed energy is reradiated in all (4π) directions. Therefore, at the time of reradiation, the angular momentum is not changed.

During the process the dust essentially loses its angular momentum and goes down spirally into the Sun. A black dust grain will completely absorb the solar radiation, however, as the Poynting-Robertson Effect does not control the motion as strongly as for reflective, white surfaces. However, as mentioned in section 4.2, dark-coloured dust particles exceed 70% of the total samples assigned to a cosmic origin.

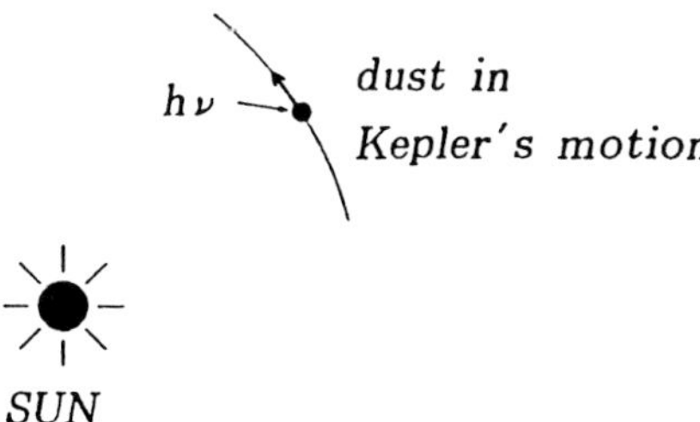

Fig. 2.1. Solar radiation controls the interplanetary dust motion due to Poynting-Robertson Effect.

This effect controls the Kepler's motion of the interplanetary dust and yields a life time, t, which is given as follows:

$$t = k(e) \cdot S \cdot \rho \cdot C \cdot r^2 \times 10^7 \quad (years),$$

where S is radius of the dust in μm, ρ is the density, r is the radial distance from the Sun in AU (astronomical units) and $k(e)$ is a constant, which is defined as the initial value of the eccentricity, e, of the orbit, such that $k(0) = 0.7$ (circular; $e = 0$), $k(0.5) = 1.9$, $k(0.9) = 7.3$ and $k(0.99) = 28.9$.

When the solar plasma flow is intense, it assists the radiation pressure against the dust motion, and $k(0) = 0.7$ changes into about $k(0) = 0.5$.

C is the energy absorption coefficient.

When the dust descends to Earth's orbit from the outer orbit, the time interval, Δt, is given by

$$\Delta t(years) = 0.7 \cdot S \cdot \rho \cdot C(r^2 - 1).$$

When a black dust grain of 100 μm in radius descends from the asteroid belt's (2.8 AU), Jupiter's (5.2 AU) and Pluto's orbit (40 AU) to Earth's orbit, we have $\Delta t = 0.24 \times 10^7$ yr, 0.9×10^7 yr and 0.56×10^9 yr, respectively.

If we capture now a "primordial" (4.6×10^9 yr before present) dust particle of 100 μm in radius, it might have started its journey at 3300 AU from the Sun!

Additional forces affecting the interplanetary dust motion are considered to be gravitational effects by planets and satellites and Lorentz interaction of the charged dust with the interplanetary magnetic field. The dust is charged positively through photoelectric effects by the solar plasma (day time) or charged negatively with electrons by the adsorption effect (dark region).

The Yarkovsky effect (Jacchia, 1963) is understood as follows: a spinning dust grain is heated by solar radiation, and the position of the most highly increased temperature on the dust surface is located at a fairly deflected angle from the center line between Sun and the dust. The revolution and the rotation of the dust take place in the same direction, so this effect causes acceleration, or if not, decleration.

However, if the spinning rate of the dust is much higher than the heat transportation time in the dust and/or if the form is very irregular, the temperature will be averaged throughout the dust grain; therefore, the heat absorption and emission processes are not as effective for the dust motion in space.

2.2 *F-Corona*

The solar corona consists of K- and F-coronas. The brightness of the K-corona accounts for more than 90% of the total and originates from high temperature plasma with a continuous spectrum. The term "K-corona" stems from the German word "Kontinuum" (continuum in English). It is said that the "F-corona" was named from the Fraunhofer corona. However, there is an objection to this, since the F-corona originates from sunlight scattered by dust rings around the Sun and the dust rings are not solar materials (indeed, almost all of the materials belong to the solar system), so "F" could be the initial letter of the German word "falsch", or the English word "false", both with the same meaning.

The K-corona is much brighter than the slightly glimmering F-corona, which can be observed only during total eclipses of the Sun.

The F-corona was independently discovered by Peterson (1967) and MacQueen (1968) by infrared observations at the time of a total solar eclipse which occurred on November 12, 1966. Their observations were performed at a wavelength of 2.2 μm and the targetted region was at 4~10 solar radii ($R_\odot$) from the Sun. In order to avoid the albedo infrared radiation from the Earth's surface, MacQueen performed a balloon observation and found bright peaks at not only 4 $R_\odot$, but also at 8.7 and 9.2 $R_\odot$. In 1970 he also reported a bright peak of 10 μm at 4 $R_\odot$. These results are shown in Figs. 2.2 and 2.3.

Maihara *et al.* (1984) confirmed the F-coronal rings experimentally at wavelengths of around 1.25, 1.65, 2.25 and 2.8 μm around 4 solar radii. They observed the surface brightness distributions at a total solar eclipse which occurred on June 11, 1983 in Indonesia by a balloon flight of 30.5 km in altitude.

Beard (1984) had already suggested that polarization measurements of the F-corona in the infrared region can give us fruitful information on the size of the circum-solar dust, so Maihara *et al.* (1984) obtained polarization at 2.25 μm and showed that the data discriminated between K- and F-corona components.

Because the scattered features of sunlight by the K- and F-coronas have no dependence on wavelength, the brightness of the K-corona was observed at each wavelength band of the photometers, so that they could extract the F-corona features as the results.

In Fig. 2.4, Maihara *et al.* showed the F-corona at 3 solar radii and compared their data with calculated values for dust sizes of 1, 10 and 100 μm. They found the spatial distribution $n(r) = r^{-1.1}$. As a result, the dust size must be as large as 100 μm or much larger.

The formation processes and important properties of the F-corona are considered to be as follows. The Poynting-Robertson Effect has given the explanation of a circular but spiral motion of interplanetary dust toward the Sun. However, the radial velocity of the dust toward the Sun begins to decrease at some distance from the Sun, because of the increase of the temperature by radiation heating and of solar radiation pressure on the sublimating dust grain.

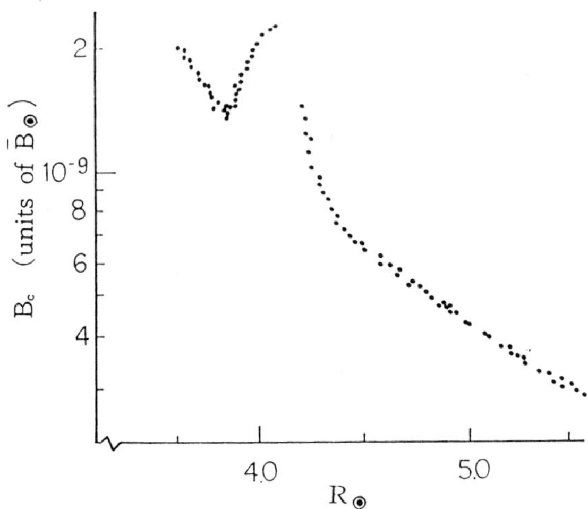

Fig. 2.2. F-corona at 4 solar radii observed by the balloon flight of Jan. 09, 1967. The ordinate represents the radiation intensity by units of those of the mean solar disc at 2.2 μm.

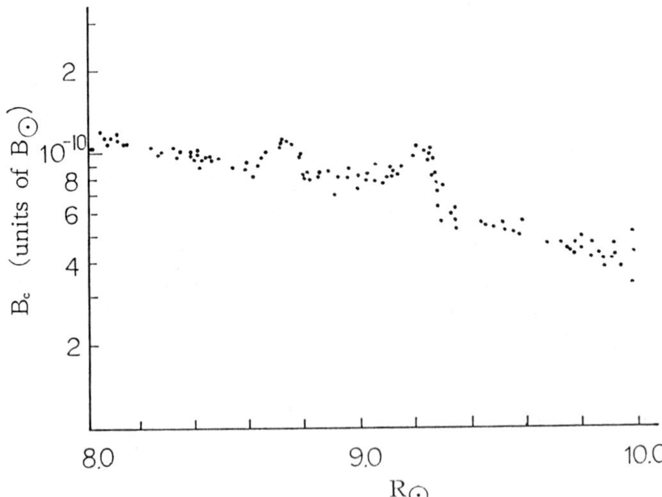

Fig. 2.3. F-coronas at 8.7 and 9.2 solar radii observed by the balloon flight of Jan. 09, 1967. The ordinate represents the radiation intensity by units of those of the mean solar disc at 2.2 μm.

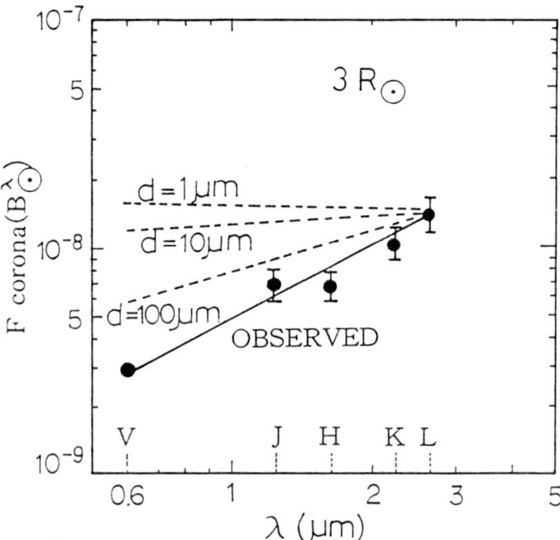

Fig. 2.4. Spectrum of the F-corona compared with calculations with parameters of various dust sizes (Haihara *et al.*, 1985).

Various solid materials contained in cosmic meteoroids have characteristic sublimation temperatures and latent heat values. Radiation pressure acts mainly on their physical factors, such as size, shape, colour, reflective properties and so on. Therefore, the increased dust temperatures by radiation heating discriminate the materials by sublimation points which correspond to the respective distances from the Sun.

Mukai and Yamamoto (1979) showed that the distance of 4 solar radii is the sublimation point as well as the point of retreat of graphite and some kinds of obsidian. They calculated the grain temperature as a function of the radial distances in units of solar radii. It is important to decide which peaks correspond to the respective materials or composite minerals in the cosmic dust.

The phenomena of F-coronas are considered to be strong evidence of the Poynting-Robertson Effect on the dust motion in interplanetary space.

Since the sublimation points are independent of the gravitational force on the dust, dust grains of different sizes begin to sublimate at nearly the same distance from the Sun.

When the radiation pressure, which depends on S^2 (S is the dust size), exceeds the gravitation force, which depends on S^3, the residual grains begin to fly away outward from the sun again.

Such a grain flux from the Sun is called β-meteoroids. The term β is introduced from the *b*-value, which is defined as the ratio of radiation pressure to the gravitational force on the dust grain. Zook and Berg (1975) called β-

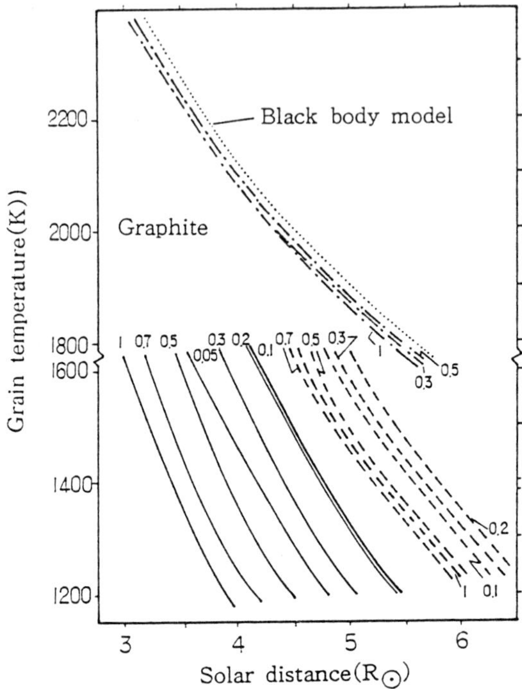

Fig. 2.5. Calculated sublimation points of graphite.

meteoroids "the grain component" which leave the solar system in hyperbolic orbits.

2.3 *Study of the Interplanetary Dust Motion by Size Dependence Determination of Cosmogenic, Long-Lived Radioisotopes*

The origin and sources of supply of the interplanetary dust have not been exactly determined; however, it is thought that the interplanetary dust moves helically toward the Sun due to the Poynting-Robertson Effect. The examination using long-lived, cosmic ray-induced radionuclides in the dust will yield fruitful information on its origins and motion, since the radioactivity of the nuclides depends on the residence time of the dust and the cosmic ray intensity in the dust's orbits. Yanagita *et al.* (1979) obtained a mean cosmic ray exposure age of the dust accreted on the Earth as 3.2×10^5 yr by determination of the cosmic ray-induced nuclides Ni-59 (7.6×10^4y) and Mn-53 (3.7×10^6y) in nickel and manganese fractions extracted chemically from deep sea sediments.

Recently, the fine cosmic dust has been captured by an aircraft or balloons in the stratosphere. However, larger samples are obtainable only from the so-called "cosmic spherules" gathered from deep sea sediments or polar ice layers. It has not yet been clearly decided whether the spherules are cosmic dust itself or ablated droplets from larger meteorites. Moreover, they suffered thermal degeneration during atmospheric entry and changed into much smaller and more rounded bodies.

In a previous work of Yamakoshi *et al.* (1982), a large number of black spherules were collected from marine sediments. Nickel, manganese, beryllium and chlorine fractions were chemically extracted for the measurement of Ni-59, Mn-53, Be-10 (1.5×10^6y) and Cl-36 (3.08×10^5y), respectively. As the first step of the work, Ni-59 and Mn-53 were measured in various spherule size groups, and the correlations between the radioactivity levels and their sizes were examined.

We picked out the black, magnetic spherules in the magnetic fractions extracted from a large volume of red clay sediments dredged at depths of more than 4500 meters off the Hawaii Islands by the research vessel Hakurei-maru of the Metal Mining Agency of Japan. After size determination, the spherules were divided into several size groups. The total weight of the spherules and the Fe, Ni, Co and Mn contents are shown in Table 2.1 (Yamakoshi *et al.*, 1982).

Table 2.1. Sample data used in the work by Yamakoshi *et al.* (1982).

SAMPLE CODE	SIZE REGION (μm)	WEIGHT (mg)	[Fe]	[Ni]	[Co]	[Mn]
FF(a)	100~140	14.9	61.7	2.7	≤0.2	1900
FF(b)	140~180	17.6	64.0	3.2	0.25	1640
F	180~260	30.5	65.1	4.1	0.31	1480
C	260~340	18.7	59.9	4.1	0.31	1100
MG	magnetic fraction including spherules	834	33.3	0.42	0.034	5380

(% by weight, Mn in ppm)

The nickel and manganese fractions were extracted chemically, and the nickel was electroplated onto disc-plates made of OFHC. The KX-rays (6.9 keV) were counted with a low-background planar Ge(Int.) (sensitive area is 500 mm^2). The counting efficiencies were determined with the values of 2π geometry, fluorescent yield, absorption losses by a Be window and polyethlene cover-sheet.

Also, the chemical yield including the electroplating process was determined using the atomic absorption method. The manganese fractions were irradiated by high flux neutrons. After purification, the induced fraction Mn-54 (312.5 d) activities were counted with an extremely low level Ge(Li) detector system.

Table 2.2. The saturation factors of Ni-59 in the comic dust.

[2.8 AU]	DUST SIZE (μm)				[5.2 AU]	DUST SIZE (μm)			
CONSTANT [k]	1	10	100	1000	CONSTANT [k]	1	10	100	1000
k = 0	0.30	0.90	1.00	1.0	k = 0	0.74	1.0	1.0	1.0
k = 1/2	0.16	0.60	0.88	1.0	k = 1/2	0.28	0.66	0.88	1.0
k = 1	0.10	0.48	0.81	1.0	k = 1	0.14	0.48	0.81	1.0

Table 2.3. Ni-59 and Mn-53 in the spherules (Yamakoshi *et al.* 1982).

SAMPLE CODE	[^{59}Ni dpm/g of spherules]	[^{53}Mn dpm/g of spherules]
FF (A + B)	0.45 (1 ± 0.20)	—
F	1.27 (1 ± 0.58)	5 (1 ± 0.6) × 10^{-3}
C	—	13 (1 ± 0.3) × 10^{-3}
M	—	—

(The statistical errors are ±σ)

Hasegawa (1978) calculated the saturation factors of Ni-59 and Mn-53 in iron dust particles with black surfaces of various sizes moving around the sun in circular orbits beyond Earth's orbit. For example, the saturation factors of Ni-59 are shown in Table 2.2. The asteroid belt (2.8 AU) and Jupiter's orbit (5.2 AU) are assumed as the starting points of the dust motion.

In these calculations, the exponents of the solar radiation flux function, $f \propto r^{-k}$, are selected as $k = 0$, 1/2 and 1.

Ni-59 has such a short half-life that the differences of the saturation factors between 2.8 and 5.2 AU can be obtained clearly only for sizes smaller than 10 μm, which is also fairly independent of the exponent *k*.

The expected values of Ni-59 and Mn-53 in secular equilibrium at 1 AU from the Sun were obtained as 1.20 and 1.88 (dpm/g of iron dust), respectively. In these calculations the exponential rigidity spectrum is used:

$$J(\geq 5MeV, 4\pi) = 150 \ (protons \,/\, cm^2, \text{sec.}),$$

where a characteristic rigidity of 100 MV and a p/α ratio of 22 were used. Ni-59 and Mn-53 activities in meteorites were, for example, 60 ± 15 and 515 ± 52 (dpm/kg) in Yardymly (iron), 12 ± 3 and 85 ± 17 (dpm/kg) in Bruderheim (stone), respectively (Honda *et al.*, 1967). Preliminary results are shown in Table 2.3.

In this work, we could find no difference in Ni-59 activities between the samples FF and F within the limits of statistical error. It is very difficult to restore the original, pre-atmospheric size of the spherules in space; however, these preliminary results are not inconsistent with Hasegawa's calculation. Ni-59 activities are much lower than the expected values as well as those in meteorites. It is well known that manganese is not so refractory, that the nickel components might be more or less lost from the cosmic iron dust grains during the heating process in the Earth's atmosphere. Therefore, quantitative estimations of the evaporation losses of Ni-59 from the dust are very difficult [see Section 7.1].

REFERENCES

Beard D. B. 1984 Astron. Astrophys. 132 317.

Hasegawa H..1978 CRL-Report 25-78-2 (Report of Workshop held in ICRR) (in Japanese)

Honda M., Arnold J. R. and Lal D. 1967 Handbuch Phys. 46/II 613.

Jacchia IL. G. and Whipple F 1950 Smithsonian Contribution Astrophys. 4 97.

MacQueen R. M. 1968 Astrophys. J. 154 1059.

Maihara T., Mizutani K., Hiromoto N., Takami H. and Hasegawa H. 1985 Properties and Interaction of interplanetary Dust ed. R. H. Giese and P. Lamy (D. Reidel) pp. 55.

Mukai T. and Yamamoto T. 1979 Publ. Astro. Soc. Japan 31 535.

Peterson A. W. 1969 Astrophys. J. 154 1059.

Robertson H P 1937 Monthly Notices Roy. Astro. Soc. 97 427.

Wyatt S. P. Jr. and Whipple F. 1950 Astrophys. J. 134 111.

Yamakoshi K., Imamura M. and Ohashi H. 1982 Proc. 5th Geochro. Cosmochrono and Isotope Geolog. (Nikko) pp. 401.

Yanagita S., Yamakoshi K. and Imamura M. 1979 Proc. Intern. Conf. Cosmic Rays (Kyoto) OG-12-20.

Zook H. A. and Berg O. E. 1975 Planet. Space Sci. 23 183.

Chapter 3

Meteor Showers, Sporadic Meteors and Cometary Dust

3.1 Meteor Showers and Their Chemical Composition

Meteor streams originating from dust clouds left behind by the orbits of comets are seen as meteor shower phenomena. For example, meteor streams derived from the periodic Halley's comet (comet P/Halley) fall on [October 3~10] in every year, when their "radiant point" is located at 335° of declination and –2° of right ascension and also on [October 18~23], when their point is at 92° of declination and +17° of right ascension. The radiant point of a meteor shower is defined as the crossing point of the Earth's orbit and the orbit of the comet which left behind the dust grains. Thus, it looks as if the meteor shower radiates from one point in the night sky.

As a famous historical event, a large-scale Leonid meteor shower occurred on November 12, 1833. In some places 2×10^5 or more meteors per hour fell. Meteor flashings often occur between 110 km and 75 km in altitude (Ionosphere-E). However, the flashing materials are not the meteor itself but evaporated, high temperature plasma trails, because the bright trails can be observed to drift in the wind of the upper atmosphere. Meteor flashing observation is, so to say, a large-scale flame-photometry in the sky. The meteor chemical compositions have been obtained as follows;

Table 3.1. The estimated chemical compositions of meteor streams.

ELEMENTS	Giacobini Stream	Perseus Stream	Leonid Stream	
	(P. M. Millman 1979)		(K. Nagasawa 1979)	
	[RELATIVE INTENSITY OR ARBITRARY SCALE]			
Fe	23.3 ± 0.5	25	28	28
Mn	—	—	—	0.01
Co	—	—	426	129
Ca	1.2 ± 0.1	5.0	0.03	0.02
Mg	1.3 ± 0.1	14.5	305	99
Na	1.3 ± 0.1	0.5	0.12	—

3.2 Sporadic Meteors and Man-Made Debris

Lindblad (1968) found that the sporadic, random meteor accretion rate changes inversely to sun-spot activity, as shown in Fig. 3.1.

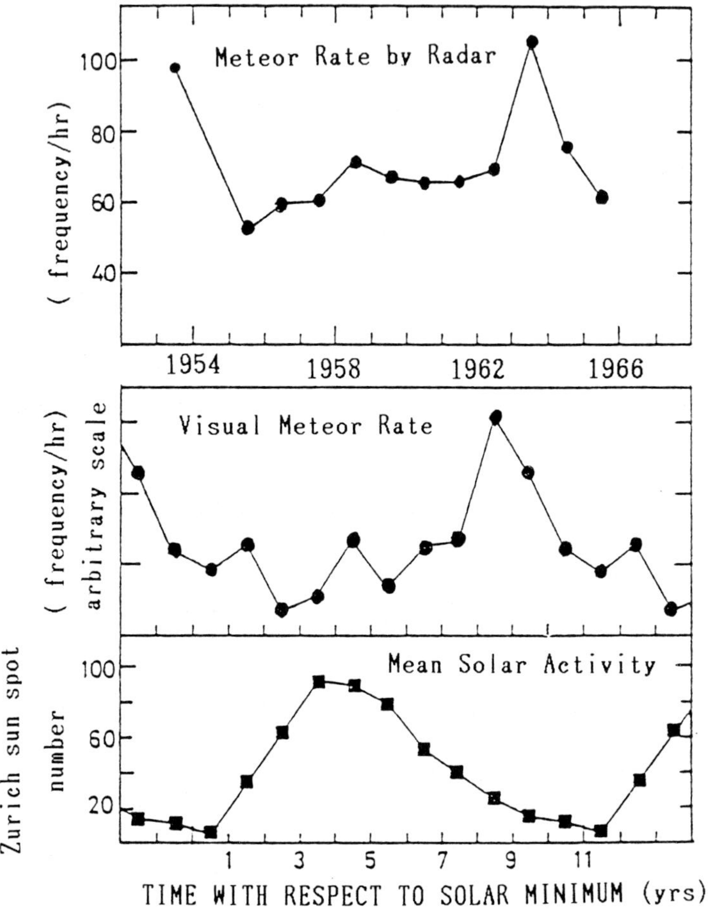

Fig. 3.1. Observed sporadic meteor rates as a function of time with respect to mean solar activity (Hughes 1976).

These data were taken in August each year. Hughes (1976) reviewed the plausible interpretations of the phenomena: (1) extraterrestrial, an actual variation in the influx of extraterrestrial dust, and (2) terrestrial, changes in the atmospheric properties affecting the probability of observing meteors. It is hardly probable that there is a physical correlation between solar activity (radiation) and meteor influx (solid grains) in the fundamental stages. Of course, there are proposed formulas, such as the Poynting-Robertson Effect etc., which is the relation between sunlight and dust motion in interplanetary space. Indeed, the periodic changes of sunlight intensity every 11 years do affect the dust motion. However, in this case there is only a relation between solar radiation intensities, variable with time, and the velocity of the dust (Kepler) motion. The 11 year cycles of dust influx over long periods should be periodic, not sporadic and random. (The Poynting-Robertson formula has no parameter for the solar radiation intensity.)

Hughes (1976) stated that the meteor trail formation process occurring at altitudes of 75~110 km could be affected by changes in the density, density gradient and scale height of the atmosphere. The density of the top layers of the Earth's atmosphere are diluted with increasing sunshine heatings, so that the observation probabilities will be decreased.

In Dr. S. Potter's presentation, "Artificial debris in high altitudes", given at a workshop held in 1989, similar phenomena were found: artificial wastes and destroyed debris accumulate and rotate the world at various altitudes; however, the mass spectrum changes with solar activity.

According to Dr. Potter, the mass spectra of the artificial debris are shown as follows:

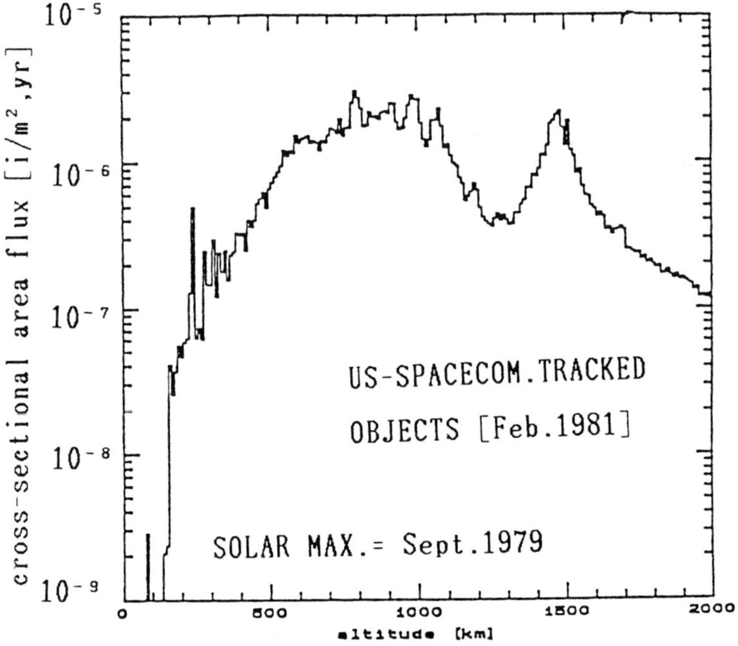

Fig. 3.2. The mass spectrum of the artificial debris (Potter, 1989).

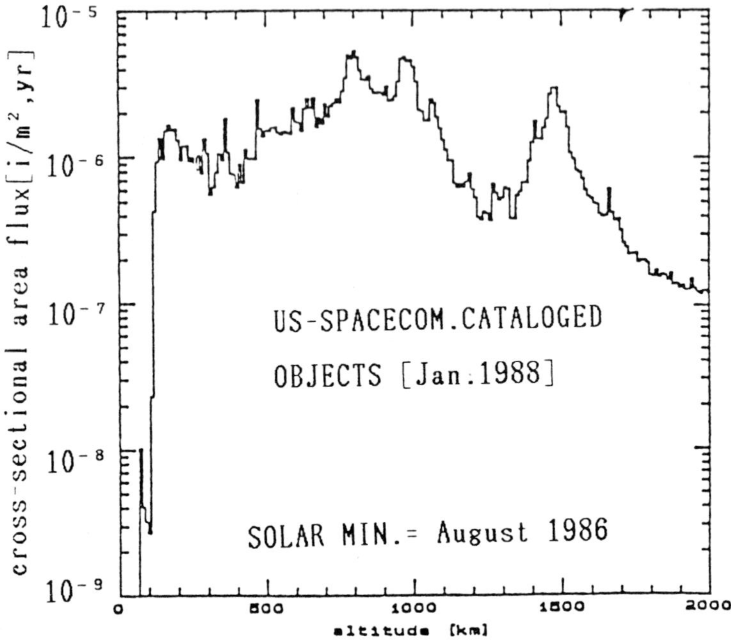

Fig. 3.2. (continued).

He interpreted this to be caused by the changes in atmospheric density due to heating by sunlight radiation. The inverse relation between the atmospheric density and solar activity can be deduced. The decreases of meteor observation frequency could be caused by the diluted atmosphere. As additional phenomena, we often observe the falling-down of artificial satellites from low altitudes onto the Earth's surface in active sun periods.

So this will be revealed clearly in the future, when the definition of "meteor observation" is precisely fixed.

3.3 Cometary Dust

It is well known that when a comet approaches the vicinity of the Sun, the sunlight irradiates the cometary nucleus and increases the temperature. The evaporated gas and plasma flow follow the comet's trajectory. Emitted dust and giant molecules are pressed in the inverse radial direction from the center of the Sun by sunlight.

From October 1985 to April 1986, a periodic comet, Halley (P/Halley) was observed by several space-borne analyzers (Soviet; Vega 1~2, ESA; Giotto, Japan; Sakigake and Suisei, NASA; ICE) as well as many ground-based telescopes.

By various space-borne detectors, not only colour pictures but also many characteristic properties and interesting events of the comet's interactions with solar radiation and magnetic fields were examined.

In the cometary dust studies of Halley's Comet, extremely fine grains of $\sim 10^{-16}$ g (subfemto grams) were discovered by in-situ measurements. They were composed of so much carbon (C), hydrogen (H), oxygen (O) and nitrogen (N) that they were called "CHON" particles (Clark *et al.*, 1987). In addition to CHON components, silicate minerals, large, mixed and exotic molecules were recognized.

By the discoveries of large amounts of CHON particles, the existence of organic materials was confirmed in comets, and the "missing" carbon problem in the solar system was understood. Moreover, the interrelation with extrasolar materials as well as the evolution of comets could be discussed.

The CHON particles are considered to be porous, just like a "fairy-castle" or birds' nest structures. In Halley's Comet new chemical species are synthesized by UV radiation triggering, and it is considered that low-temperature ice mixtures are produced by cosmic ray irradiation. The formation of complicated organic molecules were detected by sudden changes of infrared spectra. Emissions of 3.4 μm and other wavelengths suggested the presence of unknown organic materials. The cumulative mass distribution is expressed as $\sim m^{-S}$, whose power index, s, is nearly 1 at heavier than 10^{-10} g; and at 10^{-16} g, s is much lower (Fig. 3.3).

A very large polymer; $[(H_2CO_3)_5]$ = a short chain of polymerized formaldehyde, was detected in the inner coma of P/Halley with the PICCA (Positive Ion Cluster Composition Analyzer; Mitchell *et al.*, 1987) instrument on the Giotto.

The neutral gas mass spectrometer borne on the Giotto analyzed the D/H and $[^{18}O/^{16}O]$ ratios in the tail; $[^{18}O/^{16}O] = 0.0023 \pm 0.0006$, which is consistent with the terrestrial ratio = 0.00205 and the limiting range of [D/H] as $0.06 \times 10^{-4} < [D/H] < 4.8 \times 10^{-4}$, whose terrestrial ratio is $(0.0148/99.9852) = 1.480 \times 10^{-4}$. Figure 3.4 shows the [D/H] values of various planetary bodies and the samples from Halley's Comet, which are marked HNC, HCN and HCO^+ (Eberhardt *et al.*, 1987).

Isotopic ratios of $[^{12}C/^{13}C]$ are also good indicators for isotope synthesis by thermonuclear reaction. PUMA-1 analyzed the ratio and obtained no difference from the nominal value of about 90 in the solar system. $[^{12}C/^{13}C]$ ratios of Halley's Comet were obtained by ground-based spectrometers in the range 80 ± 20 (Seole *et al.*, 1987). The isotopic ratio of $[^7Li/^6Li]$ has not yet been reported.

The impact statistics of dust counters showed that the dust flux is inconsistent with the inverse-square power law of distance from the comet nucleus for a steady state and isotropic emission. Sudden increases by a factor of ~40 were often recorded. Such jet explosion type emissions of dust and gas were observed by ground-based telescopes as well.

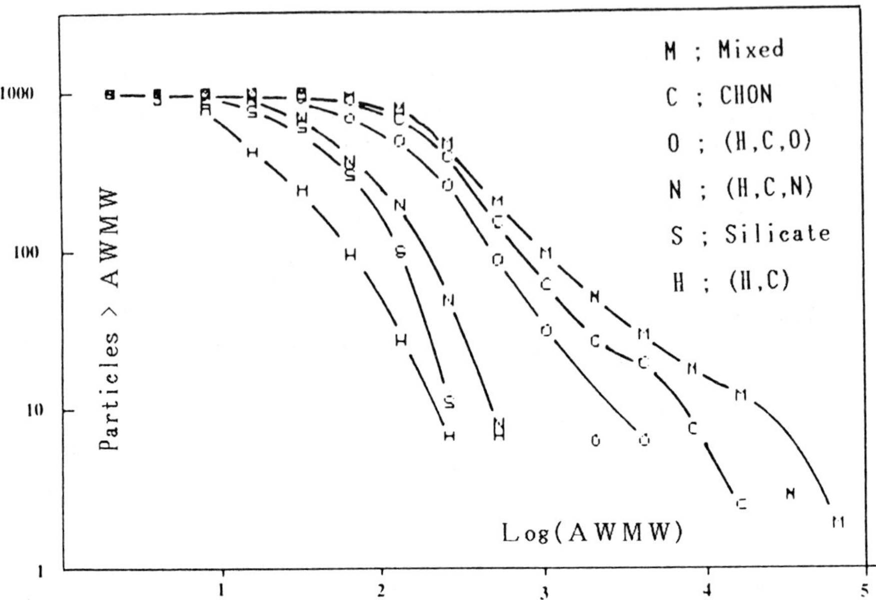

Fig. 3.3. The detected particles composed of C,H,O and N components are shown. AWMW is the amplitude representing the calculated relative mass of each grain (Clark *et al.*, 1987).

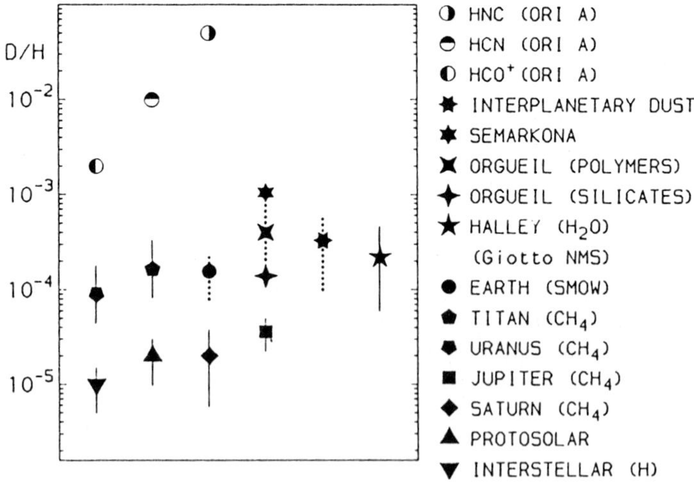

Fig. 3.4. [D/H] values in various samples (Eberhardt *et al.*, 1987).

3.4 Radiowave Reflection by Plasma Clouds Originating from Meteor Penetration in the Ionosphere

Meteor flashings originate not from the burning solid particles themselves, but from the evaporated fractions at high temperature of the incoming meteors; atmospheric trails highly heated by friction with atmospheric molecules brighten along the meteor path. Sometimes, an interesting phenomenon of the remaining bright trails drifting in the winds of the upper atmosphere can be observed.

The meteor-flashing altitude is called the E-region of the Ionosphere, in which the elcctron density increases by solar radiation in daytime, while at night the electron density decreases by one-tenth of the magnitude.

However, the electron density at night sometimes increases suddenly. Many electrons originate at the E-region of the Ionosphere. This is caused by penetrating meteors. By such electron dense clouds FM radiowaves (80~100 MHz) are reflected and can be caught for several seconds by antennas far beyond where such waves were caught usually.

Recently, the plasma clouds originating from incoming meteors have been used as one kind of "communication satellite" for data-transportation.

The data, which must be sent to far locations, are stored in the memories of devices and at the instant when meteor ionized columns are produced, the stored data are sent at a breath toward the plasma clouds. In this system, one radio station sends signals on the air continuously. When the radio signals reflected by plasma clouds arrive at another station, they act as a trigger to send the stored data to the former station. The distance between the two radio stations can be as distant as 2000 km. The "plasma-cloud communication" is being developed for military purposes in times of emergency as well, too.

C. S. Keay (1990) reported on strange sounds emitted from intense aurorae and large fire ball phenomena. These phenomena have been observed from ancient times. They are emitted from local eruptions and ionizations of air-mass along the meteor path. An ionized column due to meteor flashing grows along the meteor path, so that electromagnetic waves are generated; their wavelength can extend to the acoustic wave region.

REFERENCES

Special Issue of P/Halley; Astronom. Astrophys. (1987) vol. 187.

Clark B. C., Mason L. W. and Kissel J. 1987 Astronom. Astrophys. 187 779.

Eberhardt P., Dolder U., Shulte W., Krankowsky D., Laemmerzahl P. Hoffman J. H., Hodges R. R., Berthelier J. J. and Illiano J. M. 1987 Astronom. Astrophys. 187 435.

Hughes J. B. 1976 Space Res. XVI 333.

Keay C. S. 1990 in the talk at the Workshop on Meteor Observation held at Univ. Tokyo, Japan.

Lindblad B. A. 1968 (cited in the report by Hughes 1976).

Mitchell D. L., Lin R. P., Anderson K. A , Carlson C. W., Curtis D. W, Korth A., Reme H., Sauvaud J. A., D'Uston C. and Mendis D. A. 1987 Science 237 626.

Potter S. 1989 in the talk at the Workshop on Space Debris held at ISAS, Japan.
Seole M., Vanysek V. and Kissel J. 1987 Astronom. Astrophys. 187 385.

Chapter 4

Sampling and Analyses of Stratospheric Dust

4.1 Stratospheric Dust Collection

After considerable preparation, D. E. Brownlee and his coworkers of NASA (the Lyndon B. Johnson Space Center) have been periodically collecting stratospheric dust using an U2 aircraft since May 1981.

The collected samples are processed in an ultra-clean (class 100) laboratory. The individual particles are retrieved from the collectors, examined, classified, catalogued, and opened to dust investigators all over the world.

An entry in the Cosmic Dust Catalog is made for each dust sample retrieved from the collectors, two flat plates called "flags" (each flag has ca. 30 cm^2 surface area) coated with silicone oil.

The collectors are mounted tightly inside the wing pylons of an U2 aircraft (WB57F) and then opened at 20 km of altitude. For example, Flag-U2015 (opened in October, 1984) involved the accumulated dust samples of a total flight time of 39.6 hr.

The dust samples are gathered onto nuclear pore filters (pore size = 0.4 μm) and then attached to graphite frames (3 × 6 × 24 mm), which are mounted on a JEOL-100CX scanning transmission electron microscope (STEM). After the STEM process, the carbon-coated filters are washed with hexane to remove the silicone oil.

Each sample is examined for size, shape, transparency, colour and luster. Each page in a "Cosmic Dust Catalog" is devoted to one particle and consists of a STEM diagram and short comments and data on its size, shape, colour, transparency and luster. On the basis of the STEM results, the preliminary examination team suggests the origin of these particles: (1) cosmic origin [C], (2) terrestrial contaminants of natural [TCN], (3) terrestrial contaminants of artificials [TCA] and (4) aluminum oxide spheres [AOS].

The physical properties are described as follows:

[size (μm)]; For an irregularly shaped particle, the minimum dimension in the plane of the field of view is determined, and then the dimension at a right angle to the first one is determined. For spheres and equidimensional particles, a single dimension is measured.

[shape]; The shape of the grain is classified into three categories, [S] spherical, [E] equidimensional and [I] irregular. If the shape of the particles can not be classified as above, a description such as S/E or E/I is used.

[transparency]; The samples are divided into three categories, "transparent" (T), translucent (TL) and opaque (O) by optical microscopic investigations.

[colour]; Colour of the sample is determined by an optical microscope using fiber optic, quartz-halogen illumination and also normal reflected (tungsten lamp) illumination.

[lustre]; This parameter is determined by optical microscopy using the same illumination mentioned above. The terms (D) = dull, (M) = metallic, (SM) = submetallic, (SV) = sub-vitreous and (V) = vitreous, as well as D/SM or SV/V are used.

[type]; In these catalogs, "type" is a provisional classification, which is determined from the data of morphology (SEM) and chemical composition (XMA). Thereby the preliminary examination team (CDPET) deduces the origins of particles.

The NASA Cosmic Dust Preliminary Examination Team [CDPET] has made some short comments about the technical terms concerned with the origins of the particles; AOS, C, TCN and TCA.

[AOS]; Aluminum Oxide Spheres. AOS is transparent, subvitreous to vitreous in lustre, colourless to pale yellow and at least spherical in form. The shapes range from perfect spheres to ellipses, and the surfaces range from smooth to rough. Brownlee *et al.* (1976) found the origin of AOS to be a "space-pollutant", the product of solid fuel additive rocket exhausts. A typical figure is shown in Fig. 4.1 (NASA "Cosmic Dust Catalog" by CDPET).

[C]; Cosmic origins. CDPET deduces particles of cosmic origin to have three attributes:

(a) irregular or spherical, translucent or opaque, dark-coloured particles composed mostly of Fe with minor Ni and S,

(b) irregular or spherical, translucent or opaque, dark-coloured particles composed of various proportions of Mg, Si and Fe with traces of Al, Ca, S, and/or Ni,

(c) irregular to faceted or blocky, transparent or translucent particles composed mostly of Mg, Si and Fe, however, with traces of Al or Ca.

Categories (a) and (b) are complex, porous aggregates or distinctively "spherical" shapes and have dull to metallic luster. Chemical compositions are Fe-Ni, FeS or Fe-Ni-S phases with olivine and/or pyroxene. Category (c) is considered to be fragments of olivine or pyroxene in parent bodies. In Fig. 4.2 a typical cosmic particle is shown.

U2022 E 4

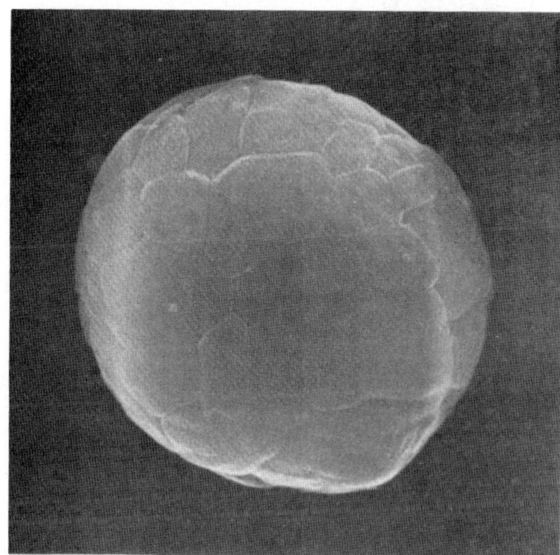

SIZE: 18
SHAPE: S
TRANS.: TL
COLOR: White
LUSTER: SV
TYPE: AOS
COMMENTS:

S-85-42825

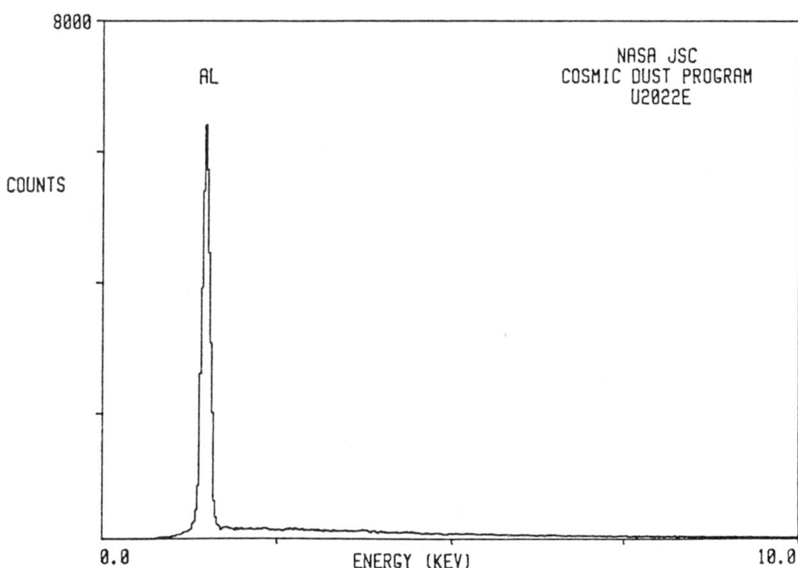

Fig. 4.1. A typical example of a grain AOS (NASA "Cosmic Dust Catalog" by CDPET).

U2022 B 1

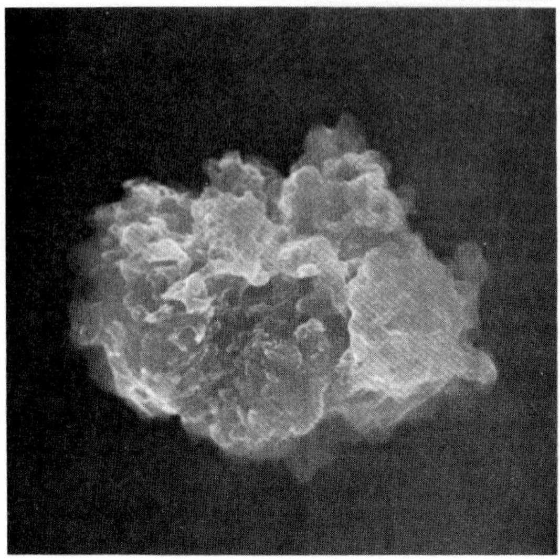

SIZE:	30x20
SHAPE:	I
TRANS.:	O
COLOR:	Brown to Black
LUSTER:	SM/D
TYPE:	C

COMMENTS:

S-85-42766

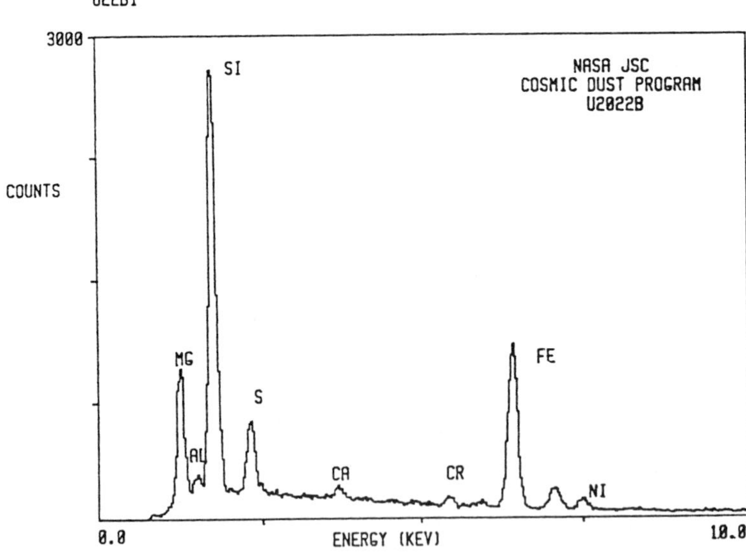

Fig. 4.2. A typical example of the category "C" (NASA "Cosmic Dust Catalog" by CDPET).

U2022 D 13

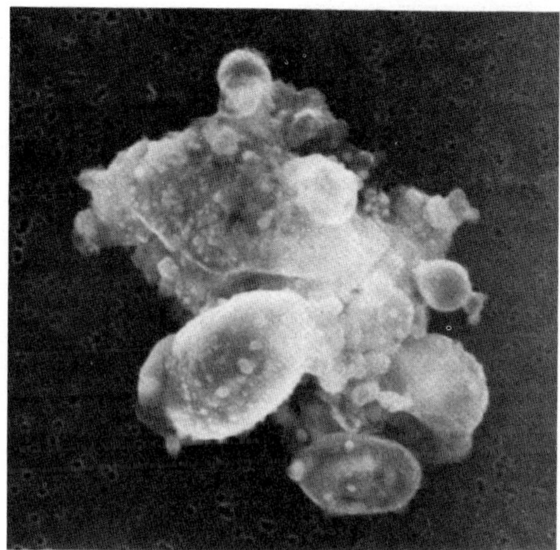

SIZE: 20
SHAPE: I
TRANS.: O
COLOR: Yellow
LUSTER: D
TYPE: TCA
COMMENTS:

S-85-42810

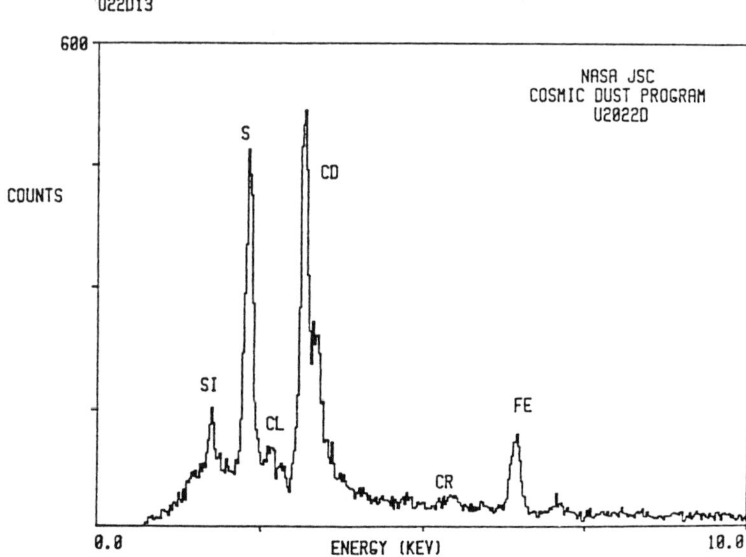

Fig. 4.3. A typical example of the category "TCA" (NASA "Cosmic Dust Catalog" by CDPET).

U2022 C 21

SIZE: 15x13
SHAPE: I
TRANS.: O
COLOR: Black
LUSTER: D
TYPE: TCN
COMMENTS:

S-85-42804

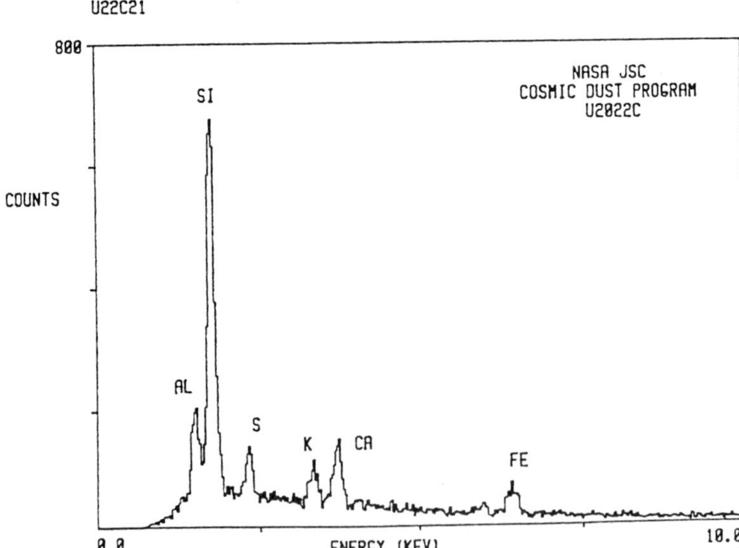

Fig. 4.4. A typical example of the category "TCN" (NASA "Cosmic Dust Catalog" by CDPET).

[TCA]; TCA particles are irregular in shape. Detected chemical elements are Al, Fe or S and as minor peaks, Ti, V, Cr, Mn, Ni, Cu, Zn, which are considered to be the components of certain metal alloys made by artificial and industrial processes. In Fig. 4.3 a typical figure of an "artificial" grain is shown (NASA "Cosmic Dust Catalog" by CDPET).

[TCN]; These are commonly irregular in shape. Chemical elements, such as Si and Al with minor abundances of Na, K, Ca or Fe are detected; these must be the components of volcanic ash in the stratosphere. For a few years after a large scale volcanic eruption has occurred, such as El Chichon, dust sampling in the stratosphere is not carried out. In Fig. 4.4 a typical figure of the category "TCN" is shown (NASA "Cosmic Dust Catalog" by CDPET).

4.2 Statistical Studies of the Specifications in the "Cosmic Dust Catalog"

Since July 1981, CDPET of NASA has periodically published the "Cosmic Dust Catalog". Nogami, Omori, Yamakoshi and Ohashi (1987) have taken the data and specifications in the "Cosmic Dust Catalog" and made a data base using a personal computer. In this section, various combinations of the data and statistical studies of the stratospheric dust will be shown. We attempt to find some suggestions for researching microfine dust samples.

As CDPET has noted, the origin classifications of C, AOS, TCN and TCA of the particles are not absolutely decisive, they only classify the collected samples for further investigations.

The input data are taken from Vol. 1 (flag; W7017) to Vol. 8 (flag; W7013).

[A] A histogram of the number of the respective categories is shown in Fig. 4.5.

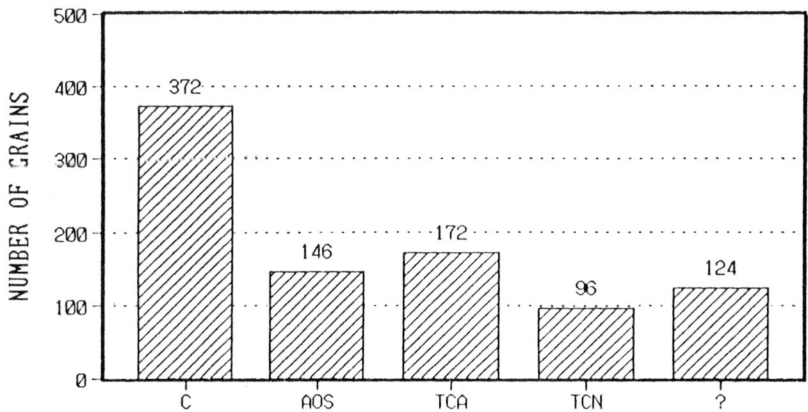

Fig. 4.5. Number of the respective cases of the stratospheric particles.

[B] Particle capture time variation of "C" and "all types" immediately after the volcanic eruptions of worldwide scale during 1980–1984, which is shown in Fig. 4.6.

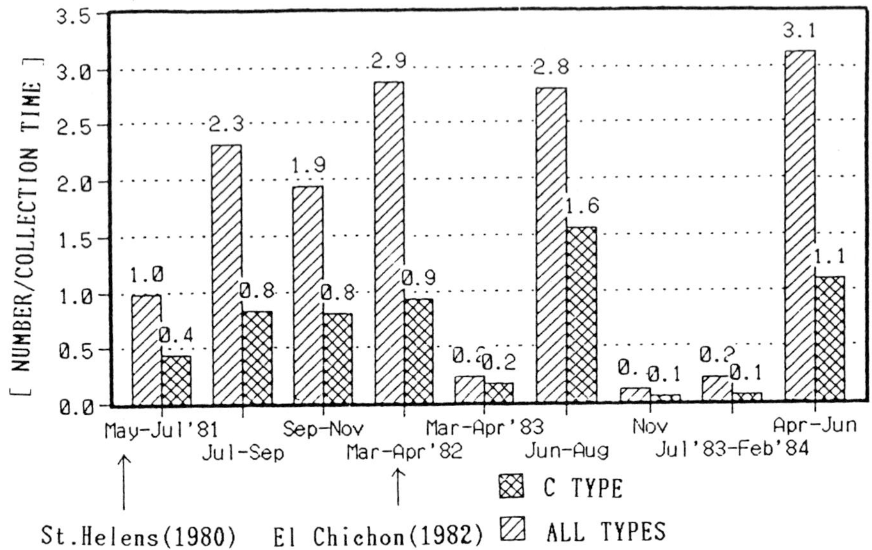

Fig. 4.6. The capture time of the stratospheric dust particles during 1981–1984.

[C] Colour properties of "C" particles are shown in Fig. 4.7. We can see that black, gray and brown coloured particles are abundant. The dark coloured particles are so abundant, that we can choose the optical absorption coefficient of the grains = 1.0, which is evaluated in Section 2.1.

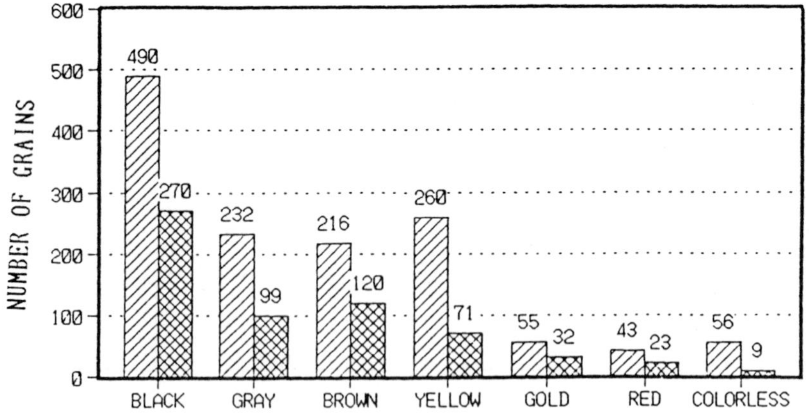

Fig. 4.7. Colour properties of the stratospheric dust.

[D] "Transparent" or "Opaque"? The "C" particles belong mainly to the "opaque" category, which is important to understand the texture of the interplanetary dust, which are shown in Fig. 4.8.

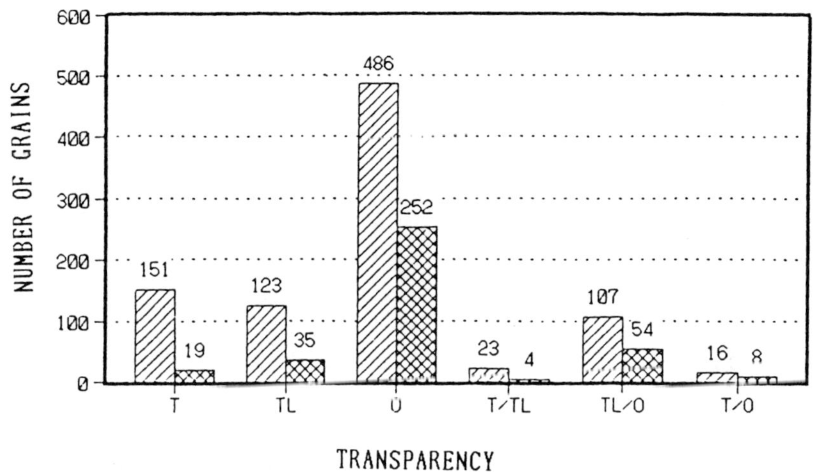

Fig. 4.8. The optical properties of the stratospheric dust.

[E] A histogram of particle forms shows that the "C" particles are abundantly irregular (S; spherical, E; equidimensional, I; irregular).

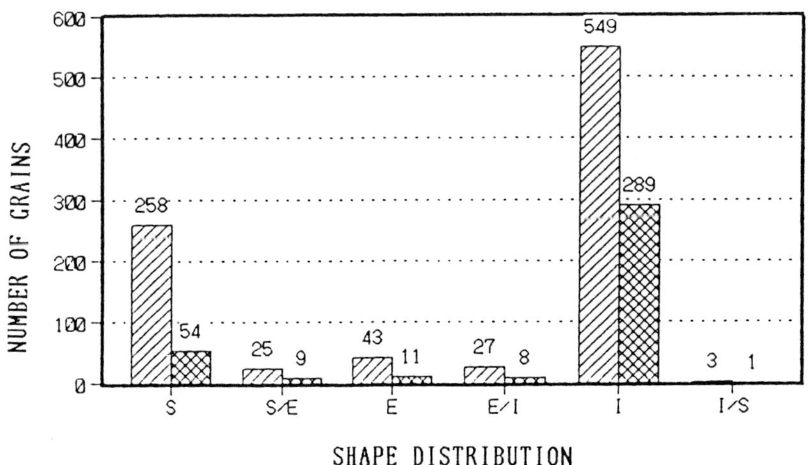

Fig. 4.9. A histogram of the shapes of the grains.

[F] A size distribution of the long and short axes can be compared with those obtained by Fujiwara (1978), who gave the ratio (a/b) ≥ 0.4 in simulation experiments.

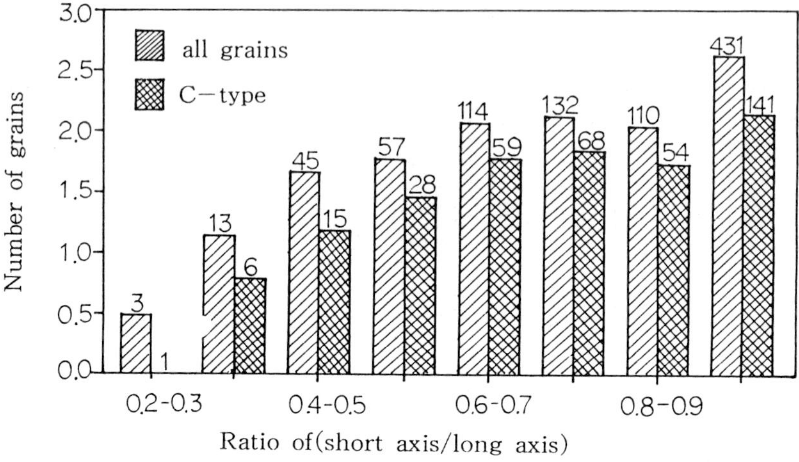

Fig. 4.10. The ratios of (short axis/long axis) of the stratospheric dust.

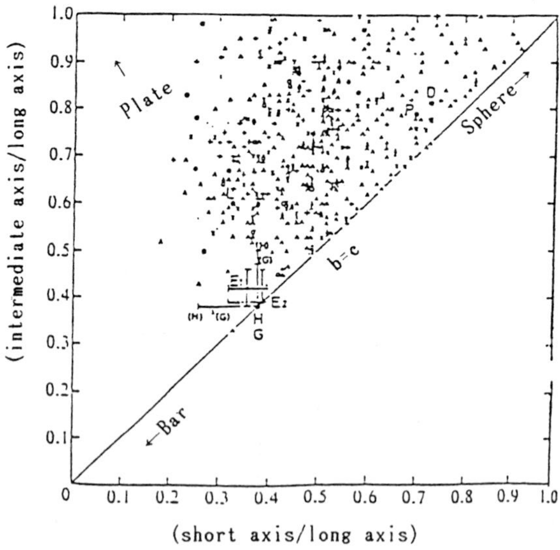

Fig. 4.11. The ratios of (short axis/long axis) and (intermediate axis/long axis) of the broken debris obtained in laboratory simulations. In Fig. 4.11, the symbol D represents the Martian satellite, Deimos, and P represents Phobos. Other English capitals on the b=c line represent asteroids (Fujiwara *et al.* 1978).

[G] A histogram of the size (long axis) of "C" and "all type" particles is shown in Fig. 4.12.

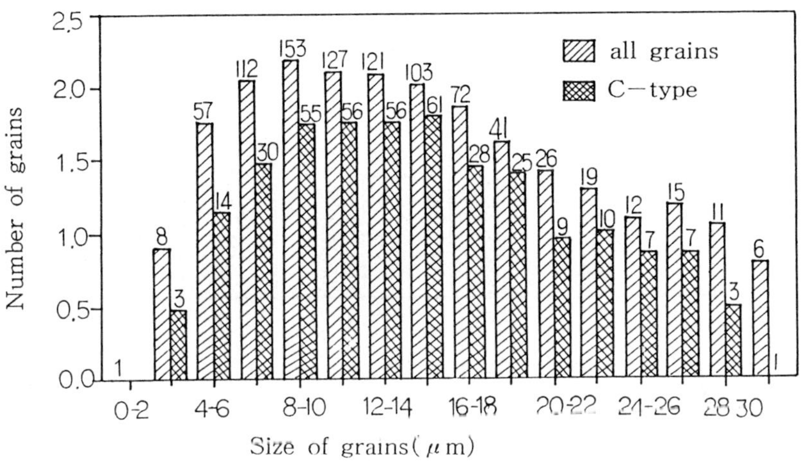

Fig. 4.12. The size distributions of "C" and "all types" are shown.

[H] A diagram of the number of particles which contain various combinations of elements; Fig. 4.13 shows siderophile compositions and Fig. 4.14 lithophile compositions.

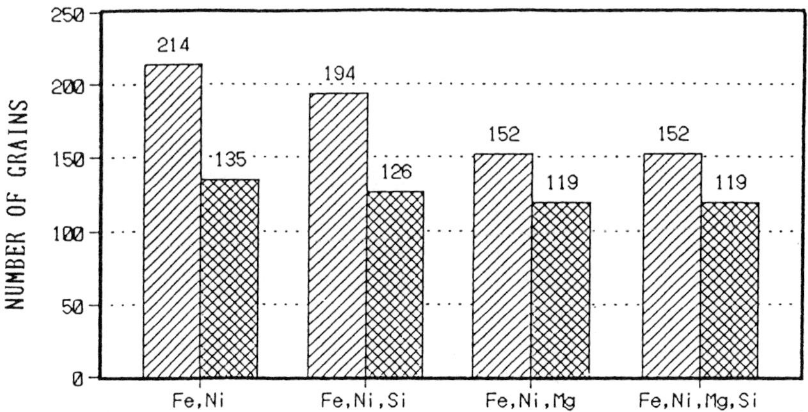

Fig. 4.13. Combinations of the siderophile compositions of "C" and "all types".

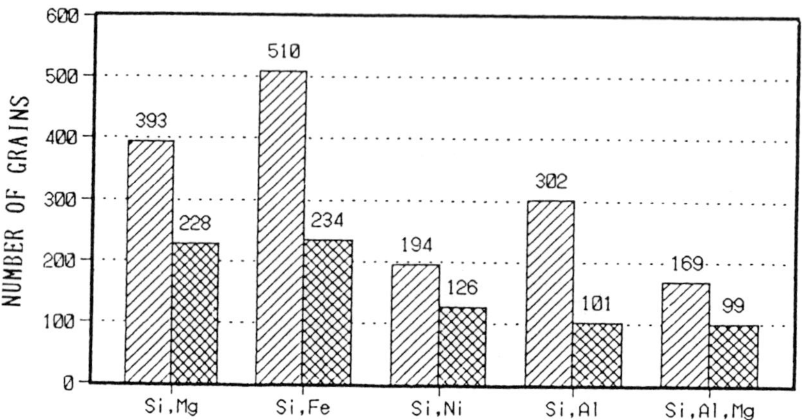

Fig. 4.14. Combinations of the lithophile compositions of "C" and "all types".

4.3 Scientific Studies on Stratospheric Dust

Any scientist can ask NASA curators to send "catalog goods", which have been listed in the "Cosmic Dust Catalog". In this section, several studies of the stratospheric dust samples will be described.

Chemical composition of stratospheric dust

Ganapathy *et al.* (1979) measured the chemical compositions of two particles using INAA. The results are shown below; the errors ($\pm 2\sigma$) are so small that they are omitted here. Both particles are black, opaque aggregates. The relative abundances of all the elements measured here, except Au, are very similar to those of C1 chondrites (Table 1).

Both particles are too small to be weighed by an ordinary microbalance, such as a CAHN/VENTRON electric balance. The mean abundances of siderophile and/or non-volatile elements, such as Fe, Ni, Cr, Co, Sc and Ir, are 7.33×10^{-8} and 10.5×10^{-8}. These figures suggest to us that the weights of these particles are 0.07 and 0.11 micrograms, respectively, because both particles have a similar composition to C1 chondrites.

It is an interesting problem that in both particles a high abundance of gold (Au) was observed. In this work of Ganapathy *et al.* (1979), it is stated that high concentrations of Au in both particles should be caused from contamination by the sample coating device. Gold is a rare metal on the Earth; however, it is also one of the most popular elements in our environment due to human activities.

In recent times, gold has been used more and more in the aero and space engineering and communication industries.

Table 4.1. The chemical compositions of the stratospheric dusts. (Ganapathy *et al.* 1979)

ELEMENTS	U2-13AI (g)	U2-14A6 (g)	C1 CHONDRITES (g/g)	ABUNDANCE RELATIVE TO C1 CHONDRITES	
				[U2-13A1]($\times 10^{-8}$ g).	[U2-14A6]($\times 10^{-8}$ g)
Fe	1.20×10^{-8}	2.28×10^{-8}	1.71×10^{-1}	7.02	13.3
Ni	9.18×10^{-10}	7.32×10^{-10}	1.03×10^{-2}	8.91	7.11
Na	4.86×10^{-10}	3.20×10^{-10}	5.11×10^{-3}	9.51	6.25
Cr	1.76×10^{-10}	3.11×10^{-10}	2.25×10^{-3}	7.84	13.8
Zn	4.89×10^{-11}	1.07×10^{-10}	3.03×10^{-4}	16.1	35.8
Co	3.82×10^{-11}	2.83×10^{-11}	4.83×10^{-4}	7.90	5.85
Sc	2.90×10^{-13}	5.68×10^{-13}	5.10×10^{-6}	5.69	11.1
Au	3.03×10^{-13}	6.39×10^{-13}	1.52×10^{-7}	199	420
Ir	3.42×10^{-14}	6.11×10^{-14}	5.14×10^{-7}	6.65	11.9

We have observed much Au in samples of smaller meteoroids. We have found a high abundance of gold in cosmic samples collected from old geological strata. This remains an unsolved problem, too.

Studies on stratospheric Al_2O_3 spheres

As described in Section 4.2, the debris collected by U-2 aircraft sometimes contained very small spherical particles of Al_2O_3 with such traces as Fe, Cu and others.

The size distribution curve is very different (much steeper) from that of the plausible extraterrestrial dust. It is shown below (Brownlee *et al.* 1976).

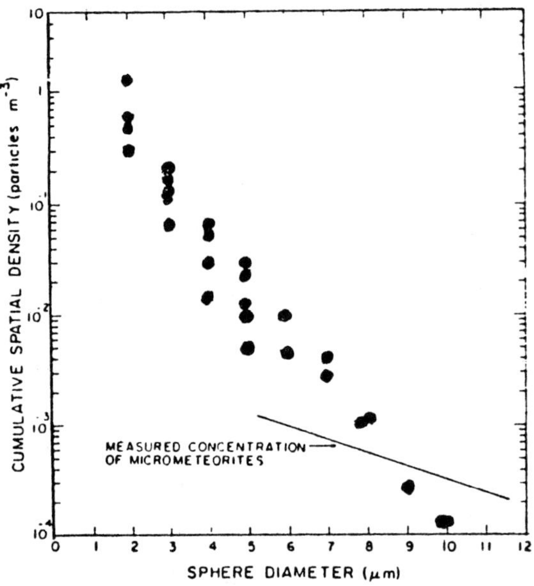

Fig. 4.15. The size distribution curve differs greatly from the cosmic micro-meteoroids (Brownlee *et al.*, 1976).

However, Brownlee *et al.* (1976) found that this type of particle is derived from solid fuel rocket exhaust. Al metal powder is often added to solid fuel to increase fuel combustion. Al_2O_3 sphere concentration can be seen as a measure of the environmental pollution in space and is also a marked disturbance to cosmic dust sampling.

Surface density and volume density determinations

Fraundorf (1982) tried to measure the mass, volume and minimum convex envelope of stratospheric dust in non-destructive forms. He used a "fish-pole balance" using a technique developed for the weighing of biological cells and tissue samples. The particle is manually picked out using a hair mounted needle and is transferred to the tip of a thin quartz fiber under a binocular microscope. The sensitivity of mass determination ranges from 2 pg ~ 1 μg. By this method, particles of 4 and 11 μm in size were weighed as 92 and 2800 pg, respectively, and the reproducibility was also very good.

The particle surface density is defined here as the ratio of mass to averaged projected area. The area and volume determinations of stratospheric dust are an important subject; they are determined by the "silhouette method". The silhouette of a particle observed from N different directions provides relative three dimensional coordinates for $N(N - 1)$ points on the minimum-volume convex envelop of the particle. Then, computer graphic techniques can reconstruct the dimensions of the indeterminate dust particle, and the area and volume can be determined with fairly large errors (Table 4.2).

Table 4.2. Surface and volume densities of the stratospheric dust samples (Fraundorf 1982).

SAMPLE	SIZE (μm)	MASS (pg)	SURFACE DENSITY (mg/cm^2)	VOLUME DENSITY (g/cm^3)	REMARKS
U13-3-02	7	690	1.1	2.2	flower
U13-3-25	7	620	0.8	1.7	cracker
U13-4-2a	11	2800	1.6	2.2	cone
U21-5-2b3	4	92	0.3	1.2	break up
U18-6-01	8	1300	1.1	2.2	wooly
U18-7-07	8	460	0.4	1.0	pancake
U23-4-4a	15	2500	0.8	0.7	spray
U21-9-01	7	1100	1.9	3.9	Al sphere
U23-4-08	8	2100	2.5	4.0	Fe^+
U23-5-09	7	770	1.0	2.0	Al prime

Detection of magnesium isotopes

Mg is one of the most interesting elements in the cosmic sciences. The magnesium element has three stable isotopes: [^{24}Mg] = 78.60%, [^{25}Mg] = 10.11% and [^{26}Mg] = 11.29%, of which ^{24}Mg is produced by nuclear synthesis in active stars. ^{26}Mg is also well known as a decay product of ^{26}Al (7.02×10^5 yr). Thus, the detection of isotopic anomalies of Mg in meteoroids has been fruitful, especially in studies of primordial meteorites and dusts. Indeed, the isotopic anomalies originate also from chemical fractionation due to heating and/or impact shocks.

Table 4.3. Mg isotopes of the stratospheric dust, lunar samples and terrestrial standards (Esat *et al.*, 1979). Here we have a formula of chemical fractionation (R) caused by known physico-chemical processes given by $R = R_0\{1 + f(M_i - M_j)\}$, where i and j are integers (mass numbers), f is a fractionation factor and dependent on temperature or impact shock. f is usually 0.01 ~ 0.05.

SAMPLE	SIZE (μm)	Mg CONTENT (10^{-9} g)	FRACTIONATION $\Delta(^{25}Mg/^{24}Mg)$ ± 2σ (per mil)	[$\delta^{26}Mg \pm 2\sigma$] (per mil)
INTERPLANETARY DUST				
MAFFIC PARTICLES				
a	22	4.5	–0.8 ± 1.4	1.7 ± 2.3
b	12	1	–1.2 ± 2.8	2.6 ± 3.1
c	23	3	–0.9 ± 0.4	–0.6 ± 0.4
d	13	0.6	–0.7 ± 1.1	0.3 ± 2.3
e	20	0.5	–0.6 ± 0.9	1.4 ± 2.0
f	10	0.1	0.3 ± 0.7	–0.5 ± 1.6
SPHERE				
*a	10	0.1	11.3 ± 0.9	0.5 ± 2.0
b	11	0.1	–2.7 ± 1.5	3.5 ± 2.5
c	16	0.3	0.3 ± 0.8	0.6 ± 1.5
CHONDRITIC AGGREGATES				
a	20	0.4	–0.8 ± 0.6	3.2 ± 1.1
b	15	0.2	2.2 ± 1.3	1.0 ± 2.8
c	30	0.8	–1.9 ± 0.5	4.0 ± 0.8
d	20	0.4	0.5 ± 0.7	3.5 ± 1.1
LUNAR ORANGE GLASS				
a	15	0.6	0.5 ± 0.9	0.0 ± 2
b	30	5	–1.7 ± 0.4	0.1 ± 0.6
LUNAR GREEN GLASS				
a	80	95	–2.0 ± 0.2	–2.0 ± 0.2
b	70	60	–0.5 ± 0.6	–1.2 ± 0.5
c	100	190	1.7 ± 0.2	–1.7 ± 0.2
d	180	150	0.1 ± 0.1	–0.6 ± 0.2
e	170	150	0.5 ± 0.2	0.2 ± 0.3
f	200	130	0.6 ± 0.2	–0.7 ± 0.2
g	210	130	1.1 ± 0.2	–0.5 ± 0.2
TERRESTRIAL OLIVINE STANDARDS				
a	20	4	0.8 ± 0.9	–0.2 ± 1.5
b	18	3	–1.0 ± 0.8	–0.4 ± 1.6
c	80	180	1.4 ± 0.1	–0.3 ± 0.2

Esat *et al.* (1979) tried to measure Mg isotopic ratios in stratospheric dust samples. Because of the extremely small sizes of the dust samples, the mass spectrometry was very difficult and demanded remarkably sensitive, tricky techniques; the results are shown in Table 4.3.

In a sphere sample (#a, ☆), a fairly large value of $\Delta(^{25}Mg/^{24}Mg)$ was obtained as 11.3 ± 0.9 (‰). However, this excess value could be considered as within the chemical fractionation factors.

In the future, successive measurements for isotopic anomalies of various elements in stratospheric dust samples will be done by more highly sensitive mass spectrometry.

Nuclear tracks in stratospheric dust

Fraundorf *et al.* (1982) have tried to make nuclear tracks in stratospheric dust samples with heavy ion accelerators. They found vague tracks in the dust, which disappeared rapidly upon electron beam irradiation by SEM. They studied the increased temperature by frictional heating and the disappearance of the tracks by heat annealing (Table 4.4).

Table 4.4. The atomic abundances of elements of the used samples compared with those of C1 chondrites. All data are normalized to silicon atoms (Fraundorf *et al.* 1982).

ELEMENT	PARTICLE [U2-20B11]	PARTICLE [U2-20B37]	C 1 CHONDRITES
Na	0.139	0.111	0.057
Mg	0.959	0.933	1.075
Al	0.064	0.044	0.085
S	0.540	0.620	0.515
Ca	0.045	0.057	0.061
Cr	0.017	0.014	0.013
Mn	0.01	0.017	0.01
Fe	0.656	0.791	0.900
Ni	0.027	0.035	0.049

In the sample U2-20-SP-57, a large (>10 μm) olivine crystal was found, and in it more than 20 nuclear tracks were discovered, which might have been caused by solar flare heavy particles. The absence of fading of the heavy tracks in the olivine suggests that it has not been heated to so high a temperature as to fade out during atmospheric entry (<600°C).

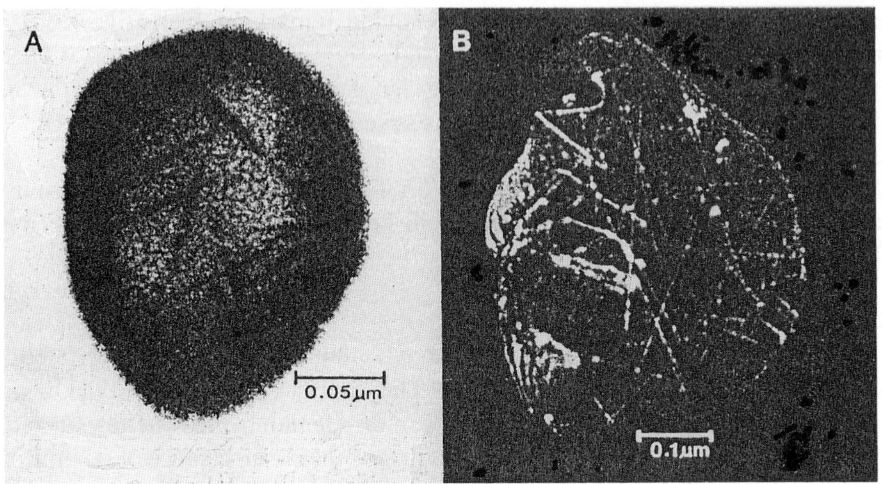

Fig. 4.16. Transmission electron micrographs are shown (Fraundorf *et al.* 1982).

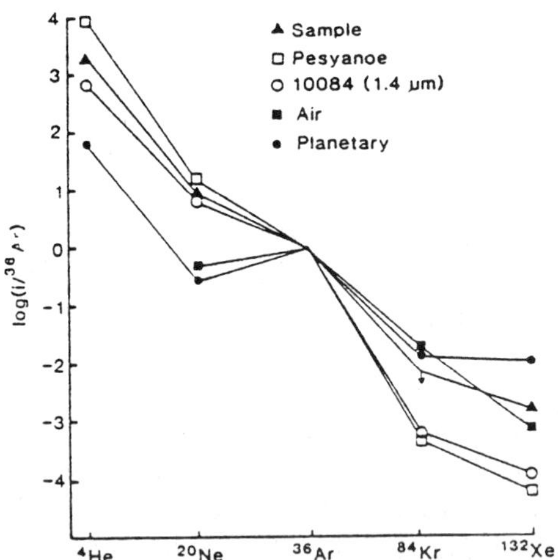

Fig. 4.17. The diagram of noble gases is shown, normalized at ^{40}Ar and compared with meteorites, lunar rocks and terrestrial air, etc. (Hudson *et al.* 1981).

Noble gases in stratospheric dust

Hudson *et al.* (1981) studied a composite sample of 13 "chondritic" dust particles and extracted noble gases, which they analyzed using an ion counting mass spectrometer. They extracted the noble gases from the composite sample by stepwise heatings at 1400, 1500 and 1600°C, dropping a nickel boat containing the dust particles into extensively degassed tungsten coils. In this work, they introduced the concept of "particle sphericity", which is defined as the surface area of a sample of equal volume divided by the particle surface area, to indicate the geometrical form of the dust particle structure (Table 4.5).

In this work, Hudson *et al.* (1981) concluded that the Ne and Ar components are caused by solar radiation; however, the Xe components might be of planetary origin. And they stated, the "apparent" Xe gases approach the contents found in noble gas-rich, acid residues in carbonaceous chondrites (Fig. 4.17).

Whiskers grown in vapour phase

Using SEM (Scanning Electron Microscope), Bradley *et al.* (1983) found pyroxene ($MgSiO_3$) whiskers, platelets and ribbons in stratospheric, chondritic porous dust samples, which are considered to be gas-to-solid condensation products. Their geometries are ca. 0.1 μm in thickness and ca. 10 μm or less in length for the rod, platelet and ribbon types. These enstatite morphologies are fragile, amorphous and highly porous aggregates, which suggests that they are vapour phase condensates in solar nebula or pre-solar circumstances.

Their high porosity shows that they might be derived from comets, and they could have survived inside cometary bodies at long, long distances from the Sun for most of their lifetime.

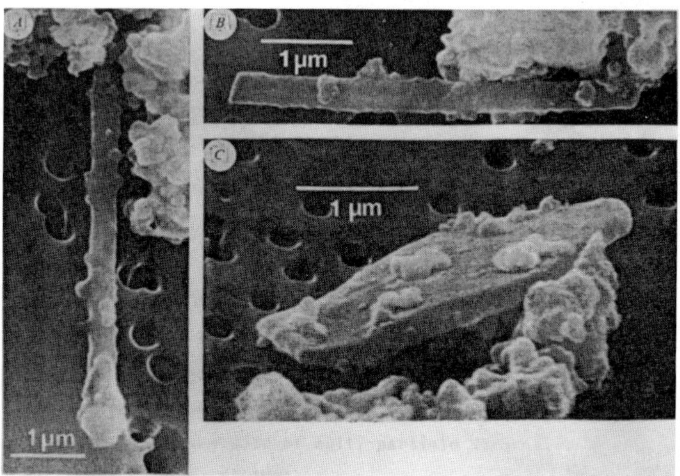

Fig. 4.18. SEM pictures, A; enstatite rod, B; ribbon and C; platelet (Bradley *et al.* 1983).

Table 4.5. The chemical compositions obtained by an energy dispersive X-ray microanalyzer and particle-sphericity (SPH) obtained by computer graphics are given (Hudson *et al.* 1981).

SAMPLE CODE	SIZE (μm)	[SPH]	[ELEMENTAL ABUNDANCES RELATIVE TO Si] Mg	Al	S	Ca	Fe	Ni
18-2-02	7.3	0.4	0.76	0.20	0.35	0.08	0.48	0.03
18-2-06	8.3	0.2	0.81	0.20	0.50	0.04	0.53	0.03
18-2-07	8.7	0.6	0.76	<0.01	0.47	0.07	0.35	0.03
18-3-01	11.4	0.2	0.62	0.18	0.23	0.05	0.52	0.06
18-3-02	14.2	0.1	0.81	0.13	0.25	0.03	0.44	0.03
18-3-07	15.5	0.2	0.67	0.10	0.57	0.02	0.51	0.06
18-4-01B	10.5	0.4	0.91	0.15	0.14	0.04	0.41	0.04
18-4-04	8.3	0.1	1.15	0.13	0.08	0.02	0.15	0.01
18-4-11	9.9	0.2	1.20	0.05	0.07	0.01	0.38	0.01
18-5-03	7.8	0.6	0.53	0.20	0.14	0.03	0.59	0.02
18-5-08	6.9	0.2	1.10	0.13	0.04	0.05	0.16	<0.01
18-5-17B	13.2	0.2	1.05	0.10	<0.01	0.03	0.33	0.03
18-5-20	19.8	0.3	0.24	<0.01	<0.01	0.01	0.60	0.03
AVERAGE	10.9	0.3	0.81	0.13	0.22	0.04	0.42	0.03

REFERENCES

Bradley, J. P., Brownlee, D. E. and Veblen, D. R., 1983 *Nature*, **301**, 473.

Bradley, J. P., Brownlee, D. E., and Fraundorf, P., 1984 *Science*, **226**, 1432.

Brownlee, D. E., Ferry, G. V., and Tomandl, D., 1976 *Science*, **191**, 1270.

CDPET team; "Cosmic Dust Catalog", compiled and published by CDPET team of NASA (Johnson Space Center).

Esat, T. M., Brownlee, D. E., Papanastassiou, D. A. and Wasserburg, G. J., 1979 *Science*, **206**, 90.

Fraundorf, P., 1976 *Geochim. Cosmochim. Acta.*, **45**, 915.

Fraundorf, P., 1982 13th Abstracts LPSC, 225.

Fujiwara, A. and Tsukamoto, A., 1978 *Nature*, **272**, 602.

Ganapathy, R. and Brownlee, D. E., 1979 *Science*, **206**, 1075.

Hudson, B., Flynn, G. J., and Shirck, J., 1981 *Science*, **211**, 383.

Nogami, K., Omori, R., Yamakoshi, K., and Ohashi, H., 1987 *Bull. Gen. Educ. Dokkyo Univ. School Medicine*, **10**, 59, (in Japanese).

Rajan, R. S., Brownlee, D. E., Tomandl, D., Hodge, P. W., Farrar, H., and Britten, R. A. 1977 *Nature*, **267**, 133.

Chapter 5

Extraterrestrial Matter Accretion into Deep-Sea Sediments

5.1 Studies on Nickel Component

In pelagic deep-sea sediments, nickel and manganese components are contained in higher concentrations than those in igneous rocks. The origin of these excess components has been discussed and examined by many workers (Pettersson and Rotschi 1950, Smales and Wiseman 1955, Laevastu and Mellis 1955, Smales, Mapper and Wood 1957, Amano, Okada and Shima 1967 and Yamakoshi and Tazawa, 1971).

Moreover, vertical distributions of nickel, iron and manganese components in deep-sea sediments were revealed by studies of long cores raised from the Swedish Deep-Sea Expedition by the Albatross in 1947~1948 (Pettersson and Rotschi, 1952).

Typical patterns of vertical distributions of the elements in a long red clay core are shown in Fig. 5. 1. The NiO content varies from a maximum of 0.089 wt% to a minimum of 0.041 wt%. Five distinct maxima occurred at various depths, and the average value was 0.056 wt%. Except for one minimum, no correlation between the variation of the three metal oxides is evident from the curves. Four sources of the nickel content in red clay were considered by Pettersson and Rotschi (1952) as follows:

An extraterrestrial supply (meteoric dust accretion) was considered to be the most probable source by them, while the high manganese concentration in red clay was considered to be supplied through volcanic matter. The interpretation of the cosmic matter supply was criticized by Smales and Wiseman (1955), Laevastu and Mellis (1955) and Smales, Mapper and Wood (1957).

Laevastu and Mellis (1955) separated magnetic spherules from the various depths in the same core which Pettersson and Rotschi (1950, 1952) studied in terms of the vertical distribution of nickel, iron and manganese components.

The size of the collected spherules ranged from 10 to 230 microns. Laevastu and Mellis (1599) obtained a relation between spherule concentration and depth in the core.

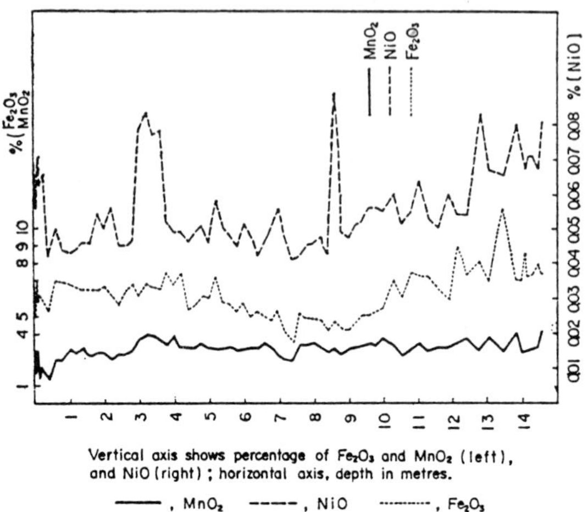

Fig. 5.1. The vertical distribution of nickel, manganese and iron contents in the sedimental core #72, obtained by Pettersson's expedition.

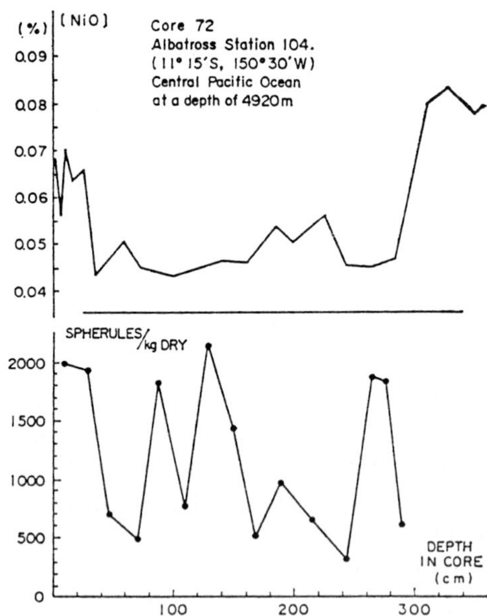

Fig. 5.2. Comparison of the vertical distribution of nickel content with metallic spherule concentration in the same core (Laevastu *et al.* 1959).

Figure 5.2 shows the concentration of magnetic spherules in the upper three meters of the core, compared with the nickel content at the same depth. They emphasized that no correlation between the spherule concentration and the nickel content in the core is present.

Smales and Wiseman (1955) pointed out that the ratios of nickel to cobalt and nickel to copper in red clay are similar to those in igneous rocks; however, they are extremely different from the ratios in meteorites. The ratios in various materials are shown in Table 5.1.

Table 5.1. Average ratios between nickel, cobalt and copper in various samples (Smales and Wiseman 1955).

SAMPLES	(Ni/Co)	(Ni/Cu)	(Cu/Co)
Red clay	3.1	0.85	3.7
Globigerina ooze	4.1	0.77	5.4
Manganese nodules	4.3	1.2	4.3
Oceanic rocks	2.3	1.6	2.7
Average for igneous rocks	3.5	1.1	3.0
Meteorites (stone and stony irons)	22.7	100	0.24
Average for meteorites	13.1	92	0.14

Moreover, the same opinion was repeated in the investigation by Smales, Mapper and Wood (1957). In that work, nickel, cobalt and copper contents in igneous rocks, marine sediments and meteorites were determined using neutron activation analyses.

Oepik also discussed this problem (1955). He stated that comparison of the ratios of nickel to cobalt and copper is not so important, and the ratio of nickel to cobalt in the solar atmosphere is rather more similar to the ratio in red clays than to the ratio in meteorites.

In 1954, Goldberg proposed an idea of the scavenging of various trace elements in sediments by the hydrated oxides of manganese and iron.

Amano, Okada and Shima (1967) separated soluble and insoluble nickel fractions by 1 N HC1 in a long deep-sea core. They considered that soluble nickel is brought as Fe-Ni alloy by cosmic material, and that insoluble nickel is supplied as a camouflage element in the silicate minerals of terrestrial origin.

From the results it can be said that the nickel component is coprecipitated on the whole with the hydrated oxides of manganese and iron into the bottom sediments. It is a noteworthy fact that similar ratios of nickel to iron and manganese components are also found in the ferro-manganese nodules, which may suggest the origin or a growth mechanism of ferro-manganese nodules in deep sea bottoms.

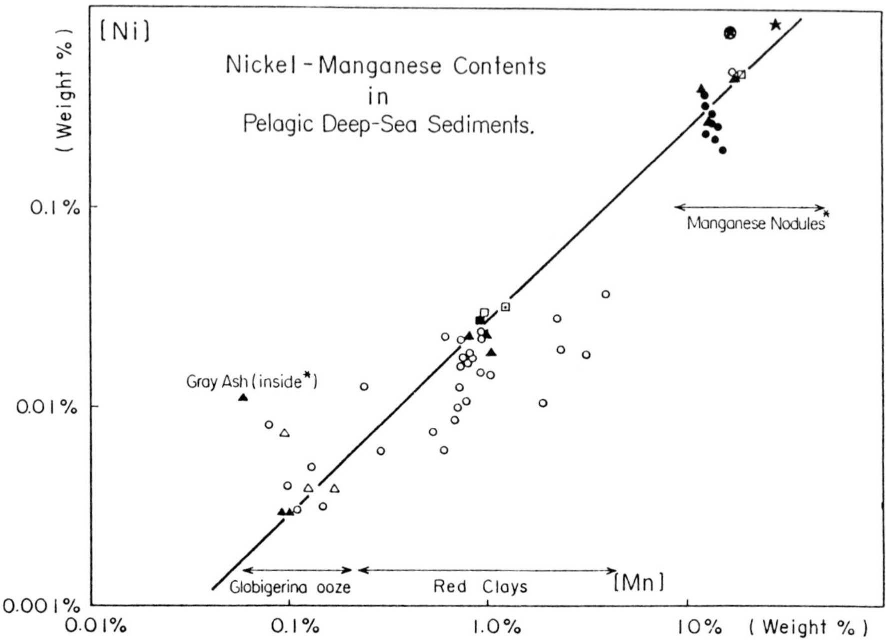

Fig. 5.3. Nickel-manganese contents in plagic deep-sea sediments.

Examination of the objections

Here I will show the positive relations between nickel, manganese and iron in deep sea red clay, globigerina ooze and ferro-manganese nodules. The data, gathered from the many articles, are shown in Figs. 5.3 and 5.4.

In the following paragraphs various objections against Pettersson's hypothesis mentioned above will be checked and refuted.

The objection by Laevastu and Mellis (1955) is not so effective against Pettersson's hypothesis; the nickel content in red clays must be compared with the total weight of the spherules, never with the number of spherules of all sizes per unit volume.

Therefore, the values of spherule number concentration limited in size from 10 to 230 microns per unit volume, which are compared with the nickel content in red clay, are just meaningless. If the total weight of spherules ranging from 10 to 230 microns in size is used, however, it may be difficult to compare them with the nickel contents in red clays, because a fair amount of nickel smaller than 10 microns is present also in red clays. Yamakoshi (1971) determined the nickel,

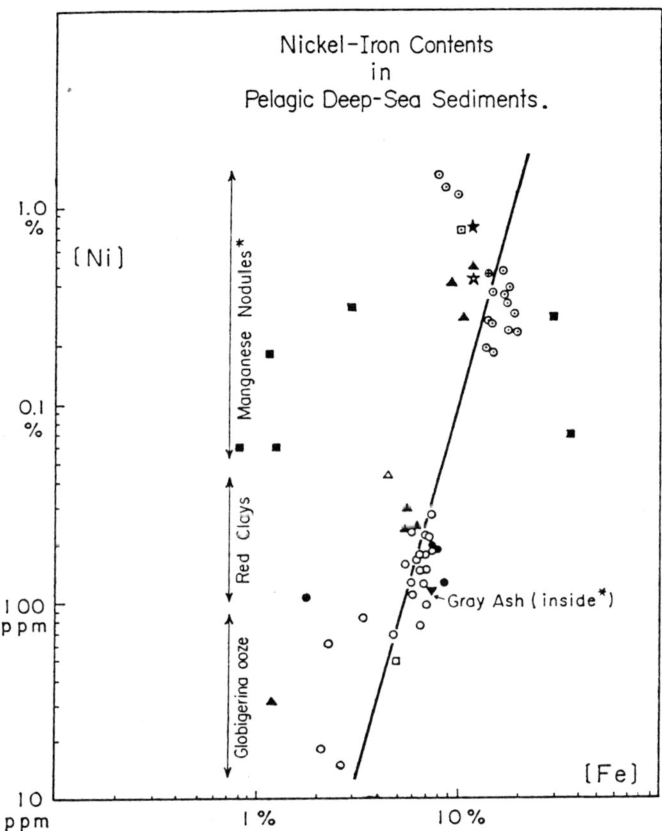

Fig. 5.4. Nickel-iron contents in pelagic deep sea sediments.

manganese and iron contents in a size fraction smaller than 6 microns in a red clay sample to be [Ni] = 0.020 wt%, [Mn] = 0.48 wt% and [Fe] = 6.28 wt%. The red clay sediment sample used was dredged at the location (0°00'N, 150°05'E) in 1968.

According to Sverdrup's data, the fraction smaller than 6 microns in size exceeds 90% of the total red clay. Therefore, in that fraction the nickel content exceeds $(0.020/0.023) \times 0.9 = 0.78$ of the total nickel in red clay, where the average nickel content is 0.023 wt% in the sample used here.

Next, the (Ni/Co) ratios obtained by Smales and his coworkers using neutron activation analysis were widely distributed from 1.4 to 7.1.

Furthermore, the average ratios of nickel to cobalt contents in sea water are estimated as 26.7 and 30.5 in the Central and Eastern Pacific Ocean, respectively; these values are larger than the ratios in meteorites. The averaged ratio of nickel to copper contents in sea water is estimated as 0.146 for 28 samples. These complicated differences may be not only cleared up by such a simple hypothesis, but also by the difference of their residence time in the oceans and so on.

Whether the supply process of the nickel component to the deep-sea sediments is by an ionic (or colloidal) precipitation and/or grain settling has been a problem. From Pettersson's curves, shown in Fig. 5.1, it can be said that the nickel fractions at the peak positions might be supplied through grains, but that the nickel fractions at the other plain positions must occur through ionic precipitation.

Is VN = constant with regard to spherules reliable?

Let us examine the reliability of an empirical relation of "cosmic spherules", VN = constant, where V is the volume of a single spherule in a given size group and N is the number of spherules in this size group. This experimental formula was proposed by Laevastu and Mellis (1955).

If no supply of terrestrial nickel to deep sea sediments has occurred and each size group of the sediments has the same size interval, the total weights of the nickel component contained in any size group are expected to be equal, because we have the modified relation ρCVN = constant in any size group, where ρ is the density and C is the nickel content in the "cosmic spherules". (Of course, the nickel content is fairly inverse-dependent on the size of the metallic spherules.)

Yamakoshi separated red clay, globigerina ooze and gray ooze into six size groups and determined the nickel, manganese and iron components in the respective size groups (Yamakoshi 1971).

At first, the size distribution of the sediment samples were obtained. These sediments were mechanically separated with sieves of various meshes in distilled water and ethyl-alcohol into six particle size groups.

Table 5.2. Size distributions of red clay, globigerina ooze and gray ooze (Yamakoshi 1971).

			(% by weight)	
SIZE GROUP (μm)	SIZE INTERVAL (μm)	RED CLAY (St-17)	GLOBIGERINA OOZE (St-25)	GRAY OOZE* (JEDS-10)
Larger than 177		0.134	7.50	6.21
177~125	52	0.298	3.14	2.20
125~74	51	0.254	3.32	5.55
74~46	28	0.456	1.68	11.19
46~20	26	2.855	10.01	38.04
Smaller than 20	20	96.003	74.35	38.82

*Gray ooze is obtained at a location off the Torishima Island (30°08'N, 139°08'E) at a depth of 1820 m.

The size distributions of red clay and globigerina ooze agree well with Sverdrup's data. Nickel, manganese and iron contents in the respective size groups were then determined; the results are shown in Tables 5.3~5.5.

Table 5.3. Nickel contents in six size groups of red clay, globigerina ooze and gray ooze (Yamakoshi 1971).

(% by weight)

SIZE GROUP (μm)	SIZE INTERVAL (μm)	RED CLAY (St-17)	GLOBIGERINA OOZE (St-25)	GRAY OOZE (JEDS-10)
Larger than 177		147	43	55
177~125	52	170	44	51
125~74	51	189	40	41
74~46	28	301	47	35
46~20	26	399	41	33
Smaller than 20	20	229	81	66

Table 5.4. Manganese contents in six size groups of red clay, globigerina ooze and gray ooze (Yamakoshi 1971).

(% by weight)

SIZE GROUP (μm)	SIZE INTERVAL (μm)	RED CLAY (St-17)	GLOBIGERINA OOZE (St-25)	GRAY OOZE (JEDS-10)
Larger than 177		0.275	0.008	0.053
177~125	52	0.510	0.017	0.240
125~74	51	1.22	0.015	0.223
74~46	28	2.13	0.037	0.169
46~20	26	2.62	0.017	0.193
Smaller than 20	20	0.698	0.045	0.302

Table 5.5. Iron contents in six size groups of red clay, globigerina ooze and gray ooze (Yamakoshi 1971).

(% by weight)

SIZE GROUP (μm)	SIZE INTERVAL (μm)	RED CLAY (St-17)	GLOBIGERINA OOZE (St-25)	GRAY OOZE (JEDS-10)
Larger than 177		3.81	0.126	3.81
177~125	52	3.69	0.139	4.54
125~74	51	3.17	0.119	4.57
74~46	28	3.29	0.159	4.27
46~20	26	3.41	0.123	3.95
Smaller than 20	20	6.29	1.53	3.28

The weights of nickel, manganese and iron components in 100-g samples are calculated in the respective size groups, which are shown in Tables 5.6~5.8. (Yamakoshi 1971).

Table 5.6. Calculated weights of nickel in six size groups per 100 g of red clay, globigerina ooze and gray ooze (mg per 100 g of sample).

SIZE GROUP (μm)	SIZE INTERVAL (μm)	RED CLAY (St-17)	GLOBIGERINA OOZE (St-25)	GRAY OOZE (JEDS-10)
Larger than 177		0.020	0.323	0.342
177~125	52	0.051	0.138	0.112
125~74	51	0.048	0.133	0.228
74~46	28	0.137	0.079	0.588
46~20	26	1.139	0.410	1.255
Smaller than 20	20	20.985	6.022	2.429

Table 5.7. Calculated weights of manganese in six size groups per 100 g of red clay, globigerina ooze and gray ooze (mg per 100 g of sample).

SIZE GROUP (μm)	SIZE INTERVAL (μm)	RED CLAY (St-17)	GLOBIGERINA OOZE (St-25)	GRAY OOZE (JEDS-10)
Larger than 177		0.37	0.60	3.29
177~125	52	1.52	0.53	5.28
125~74	51	3.10	0.50	12.38
74~46	28	9.71	0.62	18.91
46~20	26	74.80	1.70	73.42
Smaller than 20	20	656.7	33.46	111.2

In the following figures the average size, D_0, is defined as follows; the average volume, V_0, is given by $V_0 = \pi / 6 \cdot D_0^3$, assuming spherical forms.

So we have

$$V_0 (D_2 - D_1) = \int_{D_1}^{D_2} D^3 dD,$$

thus from these relations the average size,

$$D_0 = (1/2) \cdot \sqrt[3]{2(D_1 + D_2)(D_1^2 + D_2^2)}.$$

Table 5.8. Calculated weights of iron in six size groups per 100 g in red clays, globigerina ooze and gray ooze (mg per 100 g of sample).

SIZE GROUP (μm)	SIZE INTERVAL (μm)	RED CLAY (St-17)	GLOBIGERINA OOZE (St-25)	GRAY OOZE (JEDS-10)
Larger than 177		5.10	9.45	236.6
177~125	52	10.10	4.37	99.9
125~74	51	8.05	3.95	253.6
74~46	28	15.00	2.67	477.8
46~20	26	100.77	12.31	1502.6
Smaller than 20	20	6038.6	1137.6	1204.4

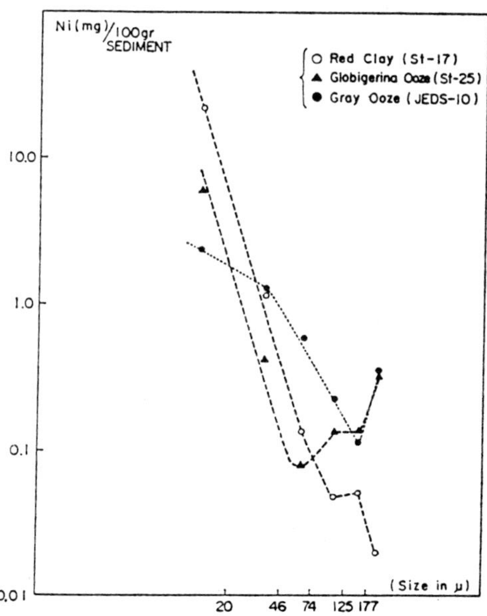

Fig. 5.5. Calculated weights of nickel in six size groups per 100 g of red clay, globigerina ooze and gray ooze (Yamakoshi 1971).

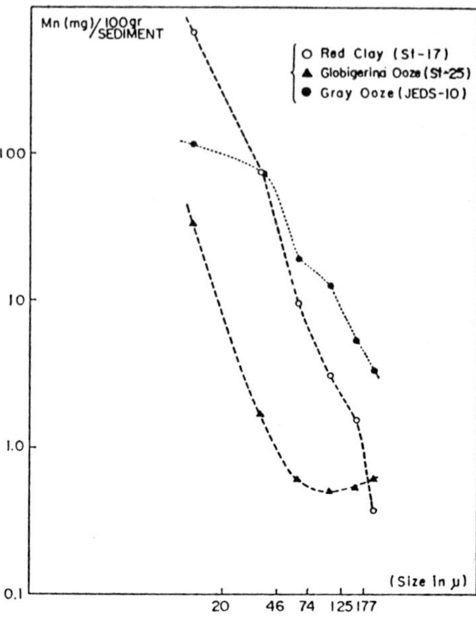

Fig. 5.6. Calculated weights of manganese in six size groups per 100 g of red clay, globigerina ooze and gray ooze (Yamakoshi 1971).

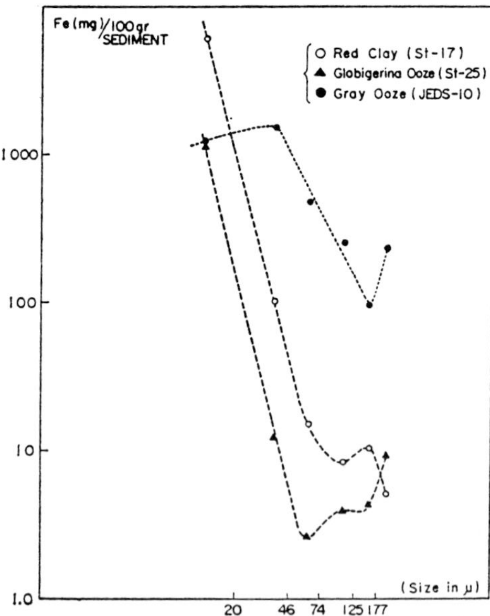

Fig. 5.7. Calculated weights of iron in six size groups per 100 g in red clays, globigerina ooze and gray ooze (Yamakoshi 1971).

The curves in globigerina ooze seem to be similar to those of red clay in a smaller size region and are quite different from the curve of gray ooze obtained off Torishima Island. The anomalous features in globigerina ooze in the larger size region may be derived from the growing of grains by calcification.

Excluding the differences of size intervals, the total weight of the nickel component is quite different in each size group. Therefore, it can be concluded, that in pelagic deep-sea sediments the nickel component supplied from grains which satisfy the relation NV = constant is not important.

Which supply of nickel to sediments occurs through grain settling or through ionic sedimentation?

Yamakoshi and Tazawa (1971) proposed a simple model calculation for nickel sedimentation into deep sea sediments.

The nickel component in pelagic deep sea sediments is considered to be supplied in two ways; grain settling and ionic (and/or colloidal) sedimentation. The former is independent of and the latter is dependent on the sedimentation rate of the sediments. The expression is shown as follows:

$$C = \frac{a + bx}{\rho x},$$

where x(cm/yr) is the sedimentation rate of deep sea deposits, ρ is the in situ density of deep sea deposits, C (g of Ni/g of sediment) is the nickel content of deep sea deposits. If a (g of Ni/cm^2, yr) is assumed to be independent of time, then a is the nickel sedimentation rate through grain settling and bx (g of Ni/cm^2, yr) is also the nickel sedimentation rate through ionic or colloidal sedimentation, in whose term it is expressed that the ionic and/or colloidal sedimentation is related to x.

A similar formulation to this model was tried by Barker and Anders (1968) in their study of the accretion rate of Ir and Os contents in deep sea sediments.

However, in that work no concept of grain or ionic supplies of the elements into deep sea sediments was expressed.

The settled grains are brought not only by extraterrestrial sources, but also by continental and/or volcanic sources.

The ionic and colloidal nickel fractions are of authigenic origin in sea water, and they are obtained almost entirely from terrestrial sources. For this model calculation, the nickel content determination of dated core samples is necessary; however, only three works have been published.

Turekian (1958) determined the nickel content in a calcareous core, Lamont A 180-76, which was dated by the radiocarbon method. The nickel concentration profile was obtained by Amano, Okada and Shima for a red clay core, Lamont V-20-130 (1967). Yamakoshi *et al.* (1972) determined mean values at two sedimentation rate regions, whose dates were determined by the paleomagnetic method. The sample code of the core used was KH-68-4, St-15, obtained by the R/V Hakuho-Maru in 1968. These results are summarized in Table 5.9.

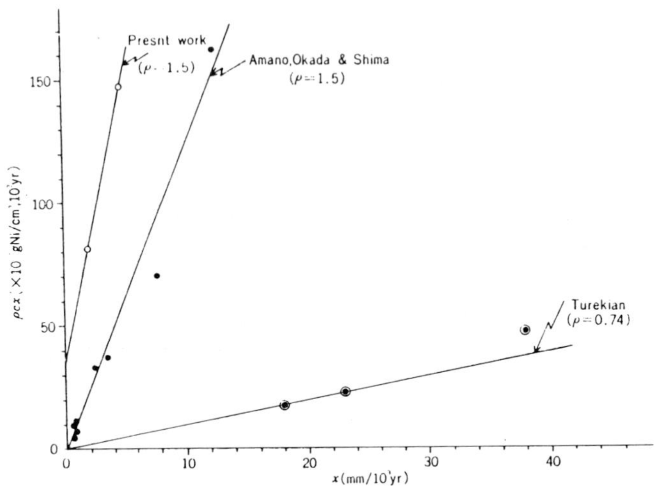

Fig. 5.8. The results are shown with the relation of ρCx vs the sedimentation rate; *x*.

The parameter a is evaluated by a least squares calculation as follows: case (a), $a = 0$; case (b), $a = 0.1$; case (c), $a = 31$ (μm Ni/cm^2, 10^3 yr), respectively. If the almost all the "grain supply" is not from terrestrial origin, we can estimate the annual accretion rate of the cosmic matter; $a = 0.1$ (μg Ni/cm^2, 10^3 yr) gives 1.5×10^4 (tons/Earth, yr), assuming the average nickel content in the "grains" to be 3.5 wt%. This is an acceptable value for us.

5.2 Studies on Iridium Components

Barker and Anders (1968) pointed out that Ir and Os contents in deep sea sediments are good indicators of cosmic matter accretion onto the Earth. They proposed that this can be used as a simple dating method, assuming a constant accretion rate of cosmic matter in the past.

However, the concept of a constant accretion rate of cosmic matter was changed suddenly to the enormous fall of an extraterrestrial body at the K-T boundary (Cretaceous-Tertiary boundary of 6.5×10^7 yr before present). Alvarez, Alvarez, Asaro and Michel (1980) published an epoch-making report. They

Table 5.9 Summarized results of nickel contents vs. sample ages.

	(a) [Core; Lamont A180-76],			(b) [Core; Lamont V-20-130],	
Zone	$\times$(cm/10^3 yr)	C (ppm)	PART	$\times$(cm/10^3 yr)	C(ppm)
A	2.3	13	A	0.76	104
B	3.8	17	B	2.47	102
C	1.8	13	C	12.5	98
	(ρ = 0.74)		D	0.81	100
			E	3.4	84
	(c) [Core; KH-68-4, St-15],		F	0.47	77
A	0.46	210	G	7.85	68
B	0.20	270	H	0.83	70
	(ρ = 1.5)			(ρ = 1.5)	

investigated many elements, especially Ir, in a Gubbio clay sample taken from the K-T boundary in Italy, and they found a very high Ir content at the K-T boundary.

They deduced that a large body (about 10 kilometers in size) crashed into the Earth and a large volume of terrestrial regolith and a part of the crushed body were ejected to high altitude and spread on a worldwide scale. The thick dust cloud covered the Earth's surface, masking it from the sun for a long time.

The moratorium of photosynthesis caused destruction of the foodchain from microscopic plants to giant animals, dinosaurs.

This scenario has been supported and/or criticized by many works and has become a hottly disputed problem at the present time. (*)

However, in the opinion of the authors, the big event at the K-T boundary was caused by a large body impact. It was not derived from an accretion event of an interplanetary dust swarm.

Alvarez *et al.* (1980) estimated an upper limit for the mass accretion rate of extraterrestrial materials as 0.024 (mg/cm^2, 10^3 yr), and Harriss *et al.* (1968) also obtained 0.024 (mg/cm^2, 10^3 yr) from studies of deep sea manganese nodules. Both works are based on the cosmic abundance of the elements.

5.3 Studies on Noble Gases in Sediments

It is well known that pelagic sediments have high ^3He contents ($\sim 10^{-11}$ cc STP/gm) as well as high $^3He/^4He$ ratios ($\sim 2 \times 10^{-4}$), which are generally much higher than those in most terrestrial samples ($\leq 5 \times 10^{-5}$).

Merrihue (1964) found high ratios of $^3He/^4He$ in magnetic fractions separated from deep sea sediments, and he deduced that the ^3He component might be caused by microfine debris of cosmic origin, which was implanted directly with solar ^3He and/or contained a cosmogenic ^3He component.

Ojima, Takayanagi, Zashu and Amari (1984) examined and concluded that almost all the sites at which ^3He anomalies are found are special regions in the ocean (such as hiatus regions) (Fig. 5.9).

However, from the vigorous examination of nearly 40 pelagic sediments collected from over 10 different locations in the Pacific and Atlantic Oceans, Ojima and his coworkers found an inverse correlation between the $^3He/^4He$ ratios and sedimentation rates. This suggests, in any case, that the samples concerned are of extraterrestrial origin.

(*) A good review of this problem was published by Lewis W. Alvarez in Physics Today. #7 (1987) pp. 24~33.

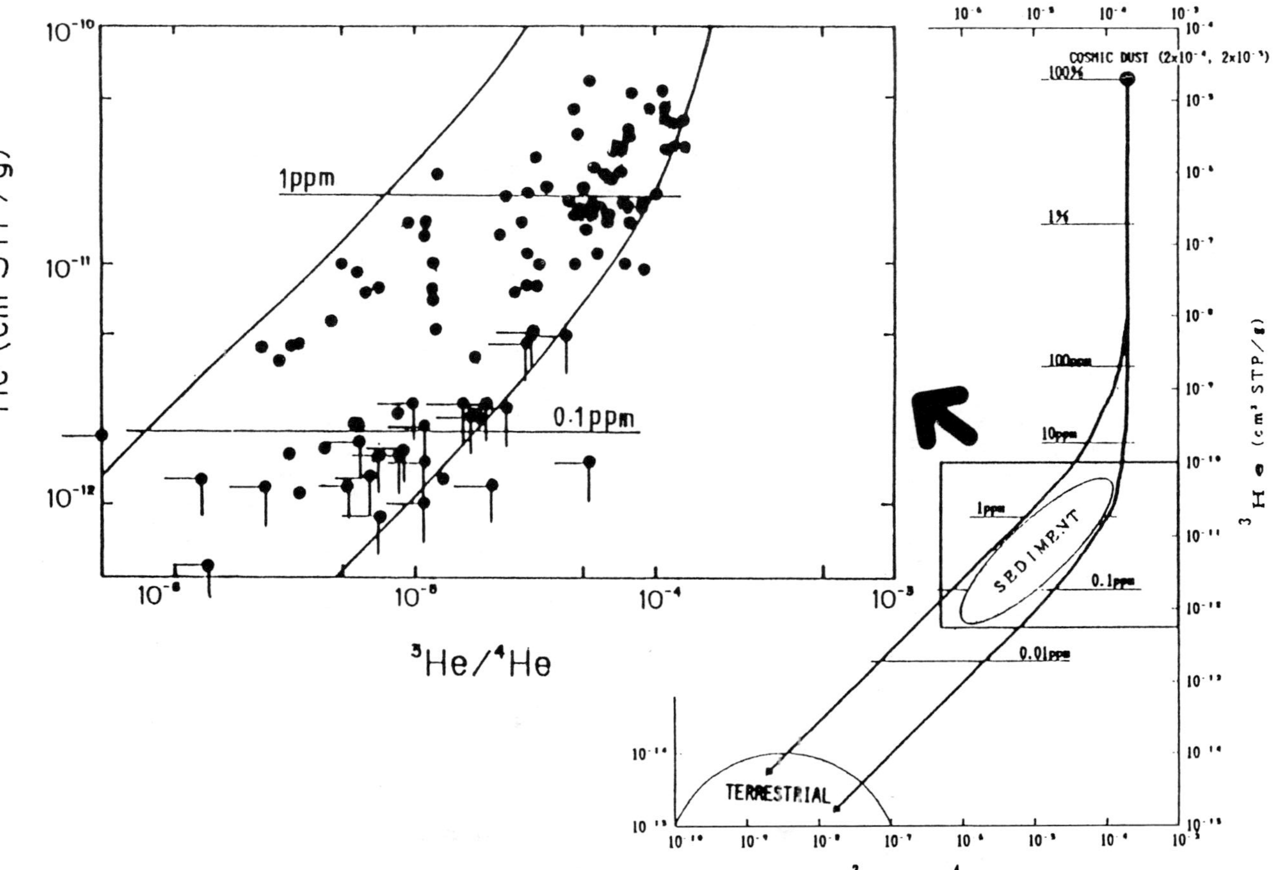

Fig. 5.9. ^{3}He contents in sediments vs. ^{3}He/^{4}He ratios are shown. The curve is drawn between the two end-members; He components in cosmic dust [^{4}He = 0.1 cc STP/gms, ^{3}He/^{4}He = 4×10^{-8}] and ^{3}He in sediments [^{3}He = 3×10^{-15} cc STP/gms, (^{3}He/^{4}He) = 10^{-8}]. All measured points are understood as mixed products of the two end members. Numbers on the curve indicate the degree of mixing of the cosmic dust in the sediments (in ppm). The various symbols in the figure show the sample locations.

Recently, isotopic anomalies in heavier noble gases than He have also been studied; much higher values [11~13] of ($^{20}Ne/^{22}Ne$) than those in air [9.8] and ($^{40}Ar/^{36}Ar$) values [<250] smaller than those in air [295.5] were also found by Amari and Ojima (1985) and by Fukumoto, Nagao and Matsuda (1986).

In a typical example of these studies, the bulk sample consisted of about 0.7 wt% magnetic fraction and 99.3 wt% non-magnetic fraction. In addition, the 0.7 wt% magnetic fraction was classified into three categories; 0.25 wt% the "strongest", 0.1 wt% "medium strong" and 0.35 wt% "weak".

The 0.7 wt% magnetic fraction in the total sediment contains over 40% of the total ^3He. The sample with the strongest magnetic intensity contained a large amount of ^3He gas; however, in these strong magnetic fractions metallic spherules are rarely found.

As a result of theoretical considerations (Hunter and Parkin 1960), a cosmic iron grain whose size is smaller than 20 μm will suffer no remarkable thermal degeneration during atmospheric entry. The critical size for chondritic meteoroids will be about 70 μm. Therefore, if we gather magnetic fractions smaller than 20 μm in size and select a portion with the strongest magnetized strength, we can expect the most abundant content of cosmic materials in it.

Time variation of ^3He contents in deep-sea sediment cores

High concentrations of ^3He as well as high ratios of ($^3He/^4He$) are observed in dredged sediments. Takayanagi and Ojima (1985) examined the time variation of ^3He and ($^3He/^4He$) ratios in sedimental cores. The inverse correlation between ^3He content and the sedimentation rate in the cores is clearly visible.

Takayanagi and Ojima (1987) calculated the flux of extraterrestrial ^3He using a simple formula;

$$\left[{}^3He\right] = f / \alpha,$$

where [^3He] (cc STP/g) is the ^3He concentration in a sample, f (cc STP/cm^2, yr) is the flux of ^3He into sediment, and α (g/cm^2, yr) is the flux of settling materials, which we call a sediment accumulation rate. Here it should be cautioned that α is different from the sedimentation rate r (cm/yr). The accumulation rate is given by a product of r and the in-situ density ρd (g/cc) which is defined as the ratio of the dry weight of sediment to the in-situ volume occupied by the sediments. Therefore we have:

$$f = \rho d r\left[{}^3He\right]$$

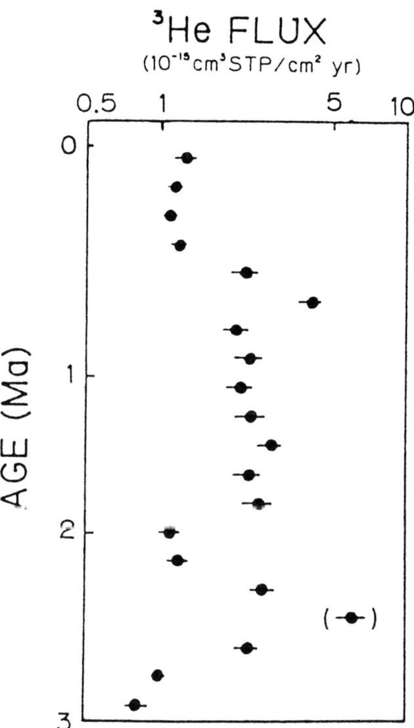

Fig. 5.10. ^{3}He content vs. age of the core (KH-68-4, St-18).*

In Fig. 5.10, we can see at around 0.7 my of the KH-68-4, St. 18 core sample anomalously high ^{3}He contents. This might be related to some local event rather than to a worldwide dust fall-out event, which is supported by Ir depth profile studies on the same core sample determined by Yamokoshi (1988).

If we consider the ^{3}He content in Brownlee particles, whose ^{3}He contents are determined by Rajan *et al.* (1977) as ^{4}He~0.1 (cc STP/g) and a solar flare ^{3}He/^{4}He ratio (= ~2) in space, the accretion rate of the interplanetary dust onto the Earth can be estimated as about 400 tons/yr. This is also an acceptable value for us.

* Sample code KH-68-4, St-18; the alphabetical code [KH] means that the sample was collected by R/V Hakuho-Maru, Ocean Research Institute, University of Tokyo, [68] means 1968, when the cruise was carried out, [4] means the 4th cruise performed in 1968, and [St-18] means the 18th point at the sampling site.

5.4 Cosmogenic Radionuclides in Sediments

In 1950, Pettersson and Rotschi noticed an anomalously large nickel content in pelagic deep-sea sediments and suggested that the excess nickel might be brought by cosmic matter into the terrestrial deep sea bottom. From the 1960's to 1970's cosmogenic long-lived radionuclides, such as ^{10}Be, ^{26}Al, ^{59}Ni and ^{53}Mn, were energetically searched for in deep-sea sediments. These nuclides were thought to be of various origins.

^{10}Be (half-life = 1.5×10^6 yr) is produced by a spallogenic nuclear reaction through N, O (p, x) in the atmosphere and/or stony and iron targets by high energy cosmic rays in space.

^{26}Al (7.02×10^5 yr) is also spallogenic through ^{40}Ar (p, x) in the atmosphere and/or stony targets with solar particles, as well as through iron targets by galactic high energy radiation in space.

^{59}Ni (7.6×10^4 yr) and ^{53}Mn (3.7×10^6 yr) are produced together from iron targets by solar particle irradiations; the former is induced predominantly by alpha particles of ^{56}Fe (alpha, n) and the latter by protons of ^{56}Fe (p, alpha). Especially, the determination of the intensities of ^{59}Ni and ^{53}Mn at the same time in dated cores is effective and useful for studying the time variation of the (H/He) emissivity of the Sun up to a few 10^5 years in the past. The ratio of (H/He) in solar cosmic rays varies in solar flares, too. Long-time solar activity can be studied by H, ^3He and ^4He measurements through noble gas analyses and [$^{59}Ni/^{53}Mn$] determinations in dated sediment core samples, and then we can investigate the results.

In these studies a large volume of red clays are chemically processed, and the expected elements are extracted quantitatively after extensive work. Sometimes these extractions are carried out with several 20-liter polyethylene buckets; thus, it has been called "bucket chemistry". The extracted species are counted by an extremely low background counter system. A series of such studies has been promoted to create and construct such low-level radiation counting techniques. In the near future, ^{59}Ni and ^{53}Mn nuclides could also be measured with particle accelerators, such as tandem Van de Graaffs or cyclotrons, by small scale sample weights.

^{10}Be in deep-sea sediments

The production channel of ^{10}Be is performed in the bombardment of the atmosphere by cosmic rays. ^{10}Be is produced, partly in the troposphere and mostly in the stratosphere. In a short time interval compared with its half-life (1.5 Ma), it reaches the Earth's surface in rain drops and settles down into deep sea sediments.

Fortunately we know a good deal information of ^{10}Be nuclide behavior, because we can follow ^{7}Be atoms as a tracer in nature, which is also produced through a spallogenic reaction in the atmosphere.

A dating method in this age region (a few million years) is well worth doing. Among events taking place during this time interval may be listed the evolution of mankind, nearly all Pleistocene glacial epochs, probably three or four reversals of the geomagnetic field and major and impulsive variations in cosmic ray intensities due to the supernovae explosions near the solar system and/or solar activity.

Our intention is to use ^{10}Be, if possible, for radioactive age determination just like ^{14}C is used in archaeology and geology. However, much work on the geochemical behavior must be done before this method can be safely used. The mechanism by which stable Be, ^{9}Be, becomes part of the pelagic sediments has yet to be elucidated. The produced atoms may be removed by adsorption or ion exchange, and/or may participate structurally in the formation of authigenic minerals.

Arrhenius (1963) furnished samples of core material which had been separated into several particle-size and density ranges by Stoke's method. The smaller grain sizes have higher Be concentrations. He suggested that the smaller grains have higher authigenic mineral contents.

Table 5.10. ^{9}Be contents in size fractions of pelagic sediments.

[particle size (μ)]	Be content (ppm)
10~32	2.1
3.2~10	2.1
1.0~3.2	2.9
less than 1.0	3.9

^{10}Be must also be involved in very small amounts in igneous rocks. The secondary reactions in the Earth which can produce it are ^{7}Li (alpha, p) and the much less important ^{9}Be (n, γ) and ^{10}B (n, p) in nature. The estimated value for the activity of ^{10}Be induced in an average granite by the first reaction is less than 10^{-3} dpm/kg, just corresponding to a ratio of ($^{10}Be/^{9}Be$) $\leq 10^{-11}$.

The production rate of ^{10}Be on a worldwide scale has been calculated by many workers: Amin *et al.* (1966) reported 1.8×10^{-2} (atoms/cm^2, sec.). Raisbeck and Yiou (1981) obtained 4.2×10^{-2}, and Reyss *et al.* (1976) calculated the global average rate as 2.1×10^{-2} (atoms/cm^2, sec.). Tanaka *et al.* (1968) estimated the value as 1.09×10^{-2} and also 4.4×10^{-2} (atoms/cm^2, sec.) from ^{10}Be analyses of core samples. These values differ quite a bit, although the collection locations of the cores were rather close to each other. Tanaka *et al.*

(1968) stated that this suggests that ^{10}Be flux depends on the differences in scavenging process, bottom topography and other marine environments. The half-life of ^{10}Be used was 1.5×10^6 yr *. Tanaka *et al.* (1986) also found a positive correlation in a core between the sedimentation rate and ^{10}Be intensities. This is also an interesting feature. It looks as if ^{10}Be atoms are derived from a terrestrial, oceanic origin.

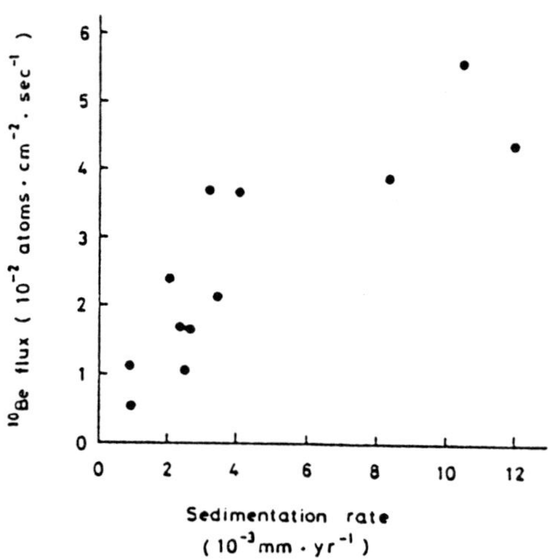

Fig. 5.11. The positive correlation between the sedimentation rate vs ^{10}Be content.

Now the coincidence of the dating of pelagic core samples by the paleomagnetic and ^{10}Be methods is so good that the ^{10}Be method could be reliable for a few million years interval by assuming constancy of the cosmic rays in the past. Recently, AMS (accelerator mass spectrometry) has improved the age determinations up to several million years before present.

Inoue and Tanaka (1982) completed a series of works on ^{10}Be intensity in several dated, deep-sea core samples collected from the Pacific Ocean. The specific concentration of ^{10}Be, $N(x)$, (in ^{10}Be atoms/cm^3), at a depth, x (cm), in a core is a function of a rate of ^{10}Be precipitation, $P(x)$, in ^{10}Be (cm/yr), as follows;

---

* In precise determinations, the values of ^{10}Be half-life are given as $(1.5 \pm 0.3) \times 10^6$ yr by Yiou and Raisbeck (1972) and as $(1.4 \pm 0.15) \times 10^6$ yr by Makino *et al.* (1975). Usually we use the value of 1.5×10^6 yr.

$$N(x) = \frac{P}{S(x)} \cdot exp[-\lambda T(x)], \ dT(x) = \frac{dx}{S(x)},$$

where λ is the decay constant of ^{10}Be (yr^{-1}), and $T(x)$ is the absolute age (yr) at a depth x. $P(x)$ and $S(x)$ may vary with time $T(x)$ and thus with x. $P(x)$ may vary with changes of cosmic ray intensity, global climate, biological productivity, oceanic circulation and other geological and cosmological conditions. $S(x)$ may vary with changes of volcanism, oceanic chemical environment, physical disturbances, biological productivity and other events occurring during the sediment accumulation time.

For the determination of the age $T(x)$, however, either $P(x)$ or $S(x)$ should be assumed to have been constant and unchanging with time during the time interval considered.

The $T(x)$ can be calculated by the following formula, assuming constant P with time;

$$T(x) = -(1/\lambda) ln\left[1 - \lambda / p \cdot \int_0^x N(x) dx\right].$$

If P is known at any location, $T(x)$ is uniquely determined for all values of x. P can be determined from a zero-depth intercept value $N(x)$. $N(0)$ and S are the mean values over the interval considered, and P can be derived by $P = N(0) \times S$. The specific activities of ^{10}Be at each layer, $N(x)$, in the core considered are plotted against ^{10}Be ages, $T(x)$, in Fig. 5.12.

The ^{10}Be concentration, in general, shows a regular decrease of ^{10}Be activity with time. The solid line on each plot represents a half-life of 1.6×10^6 yr.

The limit of ^{10}Be intensity variation from the mean life lines are ±30% for all cases, represented by dotted lines in the figure. Thus, it appears that the average ^{10}Be flux variation for these cores does not exceed ±30% for periods of 0.1 my, which is the average sampling interval. These remarkable results of ^{10}Be obtained by Tanaka and his co-workers could be succeeded with an extremely low background GM counter system ("a needle counter") (Fujita *et al.* 1975).

Lingenfelter (1979) compared two theoretical processes of the origin and acceleration mechanisms of cosmic rays in space: local sources, such as supernovae explosions (SN) and/or shock wave accelerations of diffuse and continuous media in interstellar space (ISM). He drew two diagrams typical of the expected ^{10}Be intensities in the dated cores which are shown in Fig. 5.12, too.

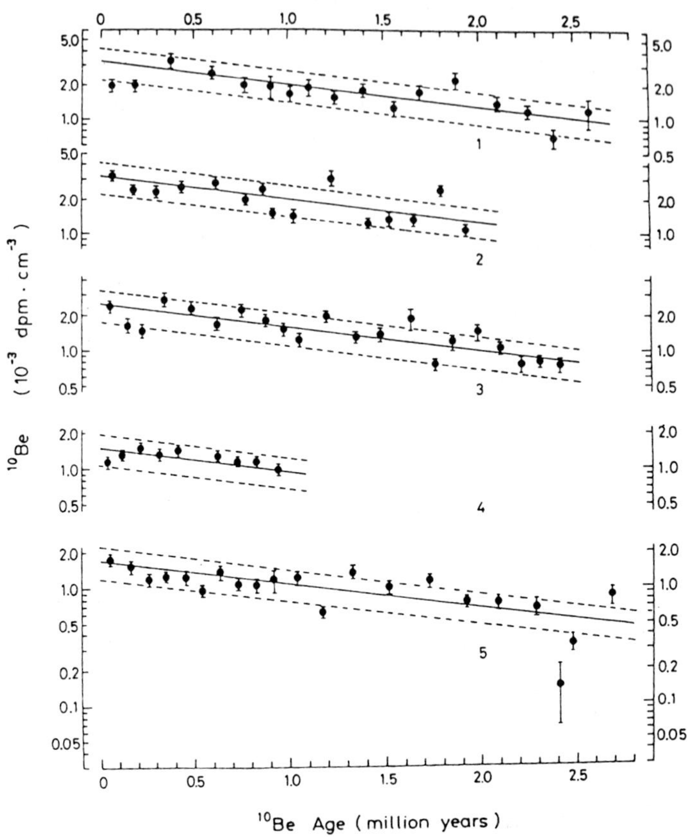

Fig. 5. 12. Comparison of the ^{10}Be concentration variations with time in five North Pacific deep-sea cores; the numbers 1–5 denote the sample cores. Mean decay lines are shown by solid lines, with ± 30% limits of variation by dotted lines.

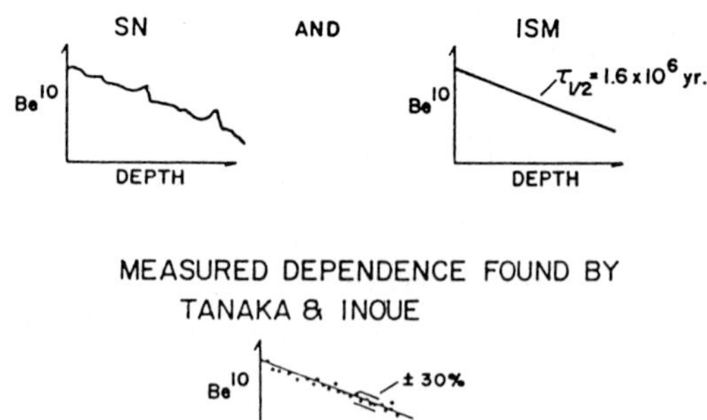

He further pointed out that the experimental evidence provided by Tanaka and Inoue (1968~1982) shown above suggested that the local and point-like origin and acceleration of cosmic rays are less preferred.

^{10}Be atoms on the Earth are produced quantitatively much more by interactions between the atmosphere and cosmic rays than the fraction of ^{10}Be brought to Earth by cosmic matter. According to Lal and Venkatavaradan (1967), the ^{10}Be atoms induced in cosmic matter in space which accrete onto the Earth are estimated as about 3.0×10^{-4} (atoms/cm^2, sec.). This is much less than the amount of ^{10}Be atoms produced in the atmosphere, by two orders of magnitude.

Here the accretion rate of the cosmic matter is assumed as 10^3 (tons/day, Earth), as is the secular equilibrium of cosmic dust influx in space. In addition, there is no evidence of such a cosmic particle accretion having ^{10}Be mentioned above.

^{26}Al in deep-sea sediments

The bulk of extraterrestrial matter accreted to Earth is in the form of micron-sized or finer grains from the interplanetary dust clouds.

A certain fraction of the dust is volatilized by interaction with Earth's atmosphere. Meteoritic fall statistics suggest that the amount of stony grains should be considerably more than that of metallic ones; however, in the deep-sea sediments stony spherules are actually so brittle that they are likely to be broken during collisions and mixing processes in the sediments. As a result, the bulk numbers of stony to metallic particles are fairly equal.

In 1963 Wasson proposed that the radioactivity inherent in the dust might allow indirect attack against some of these problems. He showed that solar flare flux should be considerably more important than cosmic rays in inducing radioactivity in the interplanetary dust. He also estimated the decay rates of several long-lived radionuclides in equilibrium that may be detectable at the Earth's surface. His calculations are based on a composition similar to that of chondrites, a solar proton spectrum similar to average spectra observed during the past solar maxima.

Two reliable sources of a large volume of accreted dust are available: the polar/Greenland ice layers and deep-sea sediments, whose advantage is isolation from sources of air-borne and water-borne terrestrial contaminants.

The ^{26}Al nuclide decays by positron emission (86%) and by γ emission of 1809 keV (96%). At that time (1960's) the gamma energy was errored as 1830 keV in almost all papers. The detector systems are composed of a gamma-gamma coincidence of positron-decayed 511 keV gammas. After high purification of the aluminum fractions of the samples, long-term measurements were carried out. As an improved system, Yamakoshi and Nogami (1975) and Nogami, Yamakoshi and Ninagawa (1977) developed a system of positron-sensitive counters and gamma-gamma coincidence spectrometer, and finally the triple coincidence beta-gamma-gamma spectrometer was completed.

Amin *et al.* (1966) extracted Be and Al fractions from Central Pacific Ocean core samples. Precise and delicate measurements were performed. Wasson (1963) estimated the disintegration rate for ^{26}Al carried by the interplanetary dust and also spallation products of the terrestrial atmospheric argon, as 1.2 and 0.08 $\times 10^{-7}$ (dpm/cm^2, yr), respectively. He stated the dust value is based on an assumed accretion rate of 10^{-7} (dpm/cm^2, yr) and assumed undersaturation of the ^{26}Al in the interplanetary dust by the factor of 10.

Tanaka *et al.* (1968) measured ^{26}Al and ^{10}Be in the samples extracted from deep-sea sediment cores. They used a 1 meter long core collected in the North Pacific Ocean from a depth of 5439 m. Al_2O_3 contents were vertically uniform as 16.38 ± 0.32 (dry wt.%) throughout the core, and values of 0.01 ± 0.13 and 4.4 ± 0.9 (dpm/dry sediment kg) were obtained for ^{26}Al and ^{10}Be, respectively.

McCorkell *et al.* (1967) processed a large volume of water (~1.2 × 10^5 liters) from 100~300 year-old ice layers at Camp Century in Greenland by columns containing cation and anion-exchange resins and milli-pore filters. After the collection some chemical components were removed and purified chemically, and the collected BeO and Al_2O_3 fractions were counted in extremely low background arrays. They stated that their obtained results (1.7 ± 0.05) × 10^{-4} (atoms/cm^2, sec) agree with the estimation of the production rate from the atmospheric argon spallation by cosmic rays 1.4 × 10^{-4} (atoms/cm^2, sec.) and they found no evidence of ^{26}Al borne by cosmic dust.

Table 5.11. Summary of the results of ^{10}Be and ^{26}Al.

INVESTIGATORS	ACTIVITY [dpm/dry sediment kg] [^{26}Al]	[^{10}Be]	RATIO [$^{26}Al/^{10}Be$]
•IN PACIFIC SEDIMENTS			
Reyss *et al.* (1976)			
at Issy Station	0.081 ± 0.046	4.6 ± 0.9	0.018 ± 0.011
at Gif Station	0.00 ± 0.14	4.6 ± 0.9	<0.03
Tanaka *et al.* (1968)	0.02 ± 0.26	4.4 ± 0.9	<0.06
Amin *et al.* (1966)	0.46 ± 0.17	3.9 ± 0.2	0.12 ± 0.04
Wasson *et al.*	0.81 ± 0.12	—	—
•IN GREENLAND ICE			
McCorkell *et al.* (1967)	0.32 ± 0.09	18.4 + 8.4 − 4.8	0.017 + 0.007 − 0.009

Reyss and Yokoyama (1976) measured the same samples used by Tanaka *et al.* (1968), and they compiled a series of results of ^{10}Be, ^{26}Al and [$^{26}Al/^{10}Be$] ratios. They concluded that [$^{26}Al/^{10}Be$] ratios obtained from two different sources (such as deep sea sediments and Greenland ice) are in good agreement, and they also agree with the expected value (0.013 ± 0.006) for the production by cosmic rays in the atmosphere. Here the contribution of cosmic dust bearing ^{26}Al can be estimated to be 0.6×10^{-9} (dpm/cm^2, yr), which is quite a bit smaller than the production of ^{26}Al in the atmosphere; 5×10^{-5} (dpm/cm^2, yr).

^{59}Ni in deep-sea sediments

Radiocarbon determination in dated samples, such as tree ring samples, has brought us much fruitful information on the solar-terrestrial connections, such as in the Maunder, Spoerer and Wolf Minima during the last 600 years and the Grand Maximum around the 11th century AD. From the last two decades we could obtain information on ^{14}C variation caused by the geomagnetic changes, as well as by solar cycles. These results have been collected by low background beta counter systems and also high sensitive accelerator mass spectrometer techniques.

In order to know the variation of cosmic rays in the distant past, one has to study radioisotopes with longer half-lives than that of ^{14}C. For that purpose we take the radioisotopes ^{59}Ni (7.6×10^4 yr) and ^{53}Mn (3.7×10^6 yr). In this section we will summarize the information on the ^{59}Ni nuclide in nature.

The variation of solar activity over a few hundred thousand years can be inferred from measurements of ^{59}Ni activity in various, dated samples.

The following processes are considered to contribute to the production of ^{59}Ni in larger bodies, such as meteorites or lunar surface rocks:

(1) spallation of ^{60}Ni, ^{63}Cu etc.,

(2) radiative capture, ^{58}Ni (n, γ),

(3) proton reaction, ^{60}Ni (p, pn), ^{59}Co (p, n) etc.

Processes (1) and (2) are induced by high energy Galactic cosmic rays and (3) by solar cosmic rays.

For smaller bodies, such as micrometeorites and interplanetary dust, the situation is quite different. In that case, the solar alpha particle induced reaction, ^{56}Fe (alpha, n), is the most dominant.

In order to determine the precise production rate of ^{59}Ni in cosmic dust, we must measure the excitation function of ^{56}Fe (alpha, n), ^{57}Fe (alpha, 2n) by cyclotron experiments (Yanagita, Yamakoshi and Gensho 1978). In addition to the imperfect data obtained by Wahlen (1969), we performed a cyclotron experiment with alpha particles whose energy ranged from 7 to 25 MeV. A Monte Carlo calculation was based on the composed nucleus model developed by Dostrovsky *et al.* (1959).

The calculations in this experiment were carried out at 1 MeV intervals, and according to Weisskopf, the dependence of the level density on the excitation energy of the excited nucleus E was assumed to be

$$W(E) = C \cdot \exp 2\left\{\sqrt{a(E-\delta)}\right\},$$

where C is a constant, δ is a pairing energy term, and a is the level density parameter.

In this calculation, a was taken as $1/20\ A = 2.8$ and $1/10\ A = 5.6\ \mathrm{MeV}^{-1}$ and the nuclear radius as $1.5\ A^{1/3}$ fm.

Figure 5.13 shows the experimental data and the calculation curves. A fair agreement is obtained with $a = 2.8\ \mathrm{MeV}^{-1}$. In these energy region, a ^{56}Fe (alpha, n) reaction seems very dominant, compared with the value through the reaction ^{57}Fe (alpha, 2n).

It is difficult to obtain the energy spectrum of solar cosmic rays in the past, so we referred to the studies of cosmogenic nuclides in lunar samples and used the intensity as $J(> 10\ \mathrm{Mev}, 4\pi) = 100$ (protons/cm^2, sec.) as well as an exponential rigidity spectrum with a characteristic rigidity of 100 MV for the solar proton flux.

The ratio of proton to alpha particle flux used is 22 ± 4 (van Hollenbecke 1975). We do not know the actual ratio; we have wanted to know it for a long time as it would show us the solar activity in the past.

According to the calculation method given by Lal and Venkatavaradan (1967), the proton induced ^{59}Ni activity can be recalculated using the above mentioned solar proton and alpha ratio; it is 0.059 (dpm/g dust). The alpha particle induced ^{59}Ni is estimated as 0.24 (dpm/g dust) using the obtained cross-section shown above. Therefore, ^{59}Ni activity determination is useful for studies of alpha particle emitting activity from the Sun.

The total activity of ^{59}Ni both by proton and alpha particles in interplanetary dust is approximately 0.30 (dpm/g dust) or 21.8 (dpm/g Ni in dust). In these calculations it is assumed that the ^{59}Ni activity in the interplanetary dust is in secular equilibrium at 1.0 AU [Astronomical Unit] from the Sun. We have here assumed 25% Fe, 1.35% Ni and 520 ppm Co as the chemical abundances in the interplanetary dust.

With these values, the contribution of accreted extraterrestrial dust to the deep-sea sediment is not more than 0.02%, and the extraterrestrial nickel fraction does not exceed 1.2% of the total nickel component in deep-sea sediments.

The sedimentation rate at this dredged station is uncertain; however, 1.8 km distant from the site a core sample (KH-67-5, St.17) was obtained, and the sedimentation rate in the upper part of 4 mm/1000 yr was measured using the excess ^{230}Th and ^{231}Pa method by Uzuyama (1975). Supposing that the dredged sample represents a depth of 10 cm from the sediment surface, a simple calculation using the sedimentation of 4 mm/1000 yr gives a mean activity of ^{59}Ni in the sediment as $0.89I_0$ where I_0 is the surface value.

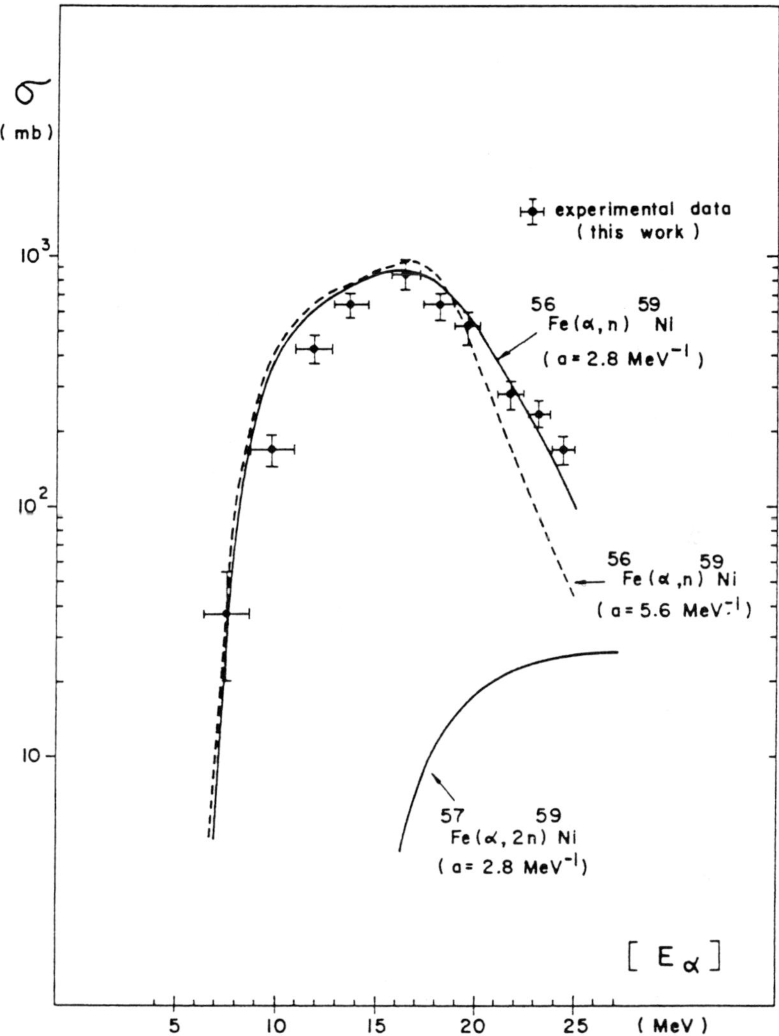

Fig. 5.13. The experimental data and Monte Carlo calculation of the excitation functions of ^{56}Fe (alpha, n) and ^{57}Fe (alpha, 2n).

Yamakoshi and Yanagita (1981) have detected the ^{59}Ni activity from red clay sediments. The sample code is KH-67-5, St.17, dredged from a depth of 5245 m, (0°01.3'S, 150°04.2'E) at the Central Pacific Ocean, obtained by R/V Hakuho-Maru, Ocean Research Institute, University of Tokyo.

The nickel content was 230 ppm and [Fe]=6.01%, [Co] = 411 ppm and [Si] = 23.02%.

From 3.14 kg of sediment, 455 mg of nickel could be extracted, which was electroplated onto a 312 cm^2 copper plate. The copper plates formed the inner wall of a cylindrical proportional counter with heavy shielding materials and a guard counter system. ^{59}Ni is an EC type nuclide, which emits KX-rays of 6.9 KeV.

The total counting time was 80495 minutes, and the blank test was performed for 38125 minutes. The energy spectra are shown in Fig. 5. 14. Just around 6.9 KeV in the energy spectrum we have an appreciable peak, and the blank curve has no peak there; the peak around 6.9 KeV could be due to the ^{59}Ni activity. The net area was converted to the absolute activity as 5.9×10^{-2} (dpm/kg sediment). The statistical error is estimated as ±30%.

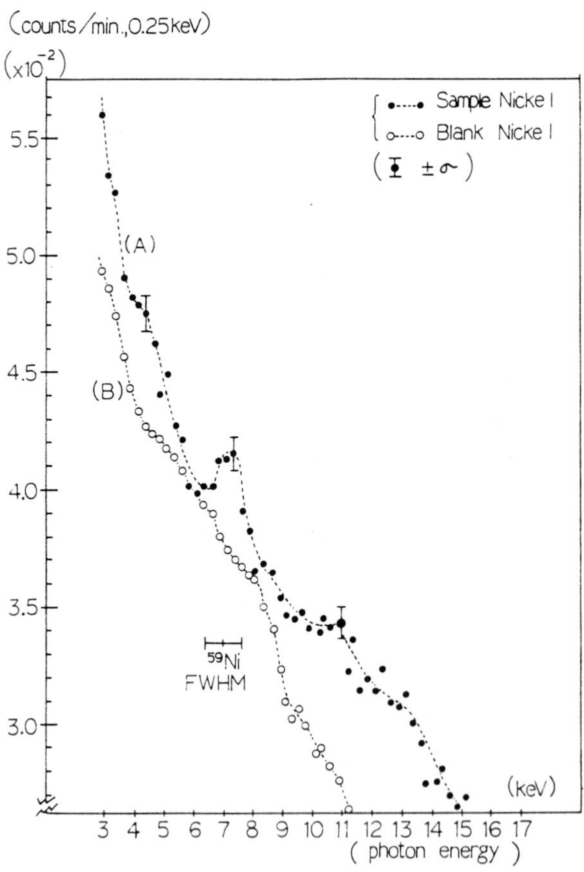

Fig. 5.14. The observed spectra of the sample-nickel (A) and a reagent-nickel (B).

In these studies some checks for contamination of artificial ^{59}Ni are necessary. After the World War II, nuclear weapons and nuclear power plants supplied much radioactive, artificial waste to the deep sea bottoms.

In the artificial wastes ^{59}Ni is induced through fission of U fuels and also neutron induced from ^{58}Ni. Therefore, we prepared 68 g of the same powdered dried sediment which was counted for 91435 minutes using a low background gamma ray spectrometer and tried to observe ^{137}Cs. ^{137}Cs is produced with such a high yield that it is a useful indicator for the estimation of artificial isotope production. The ^{137}Cs activity in the sediment was measured to be less than 0.1 (dpm/kg sediment) or 0.2×10^7 (atoms/kg sediment). The ratio of the fission yield of the mass 59 isobar to that of ^{137}Cs is expected to be less than 10^{-11}, and the expected fissiogenic ^{59}Ni contribution would be less than 10^{-13} (dpm/kg sediment).

Next, the contribution of the neutron induced ^{59}Ni was estimated. Since ^{59}Ni is mainly produced through the ^{58}Ni (n, γ) reaction, ^{60}Co was chosen as an indicator, which is produced mainly through the analogous ^{59}Co (n, γ) reaction. If the ratio of Ni and Co contents in nuclear bomb materials and wastes caused from atomic power plants is assumed as [Ni/Co] = 10, the ratio of the produced ^{59}Ni and ^{60}Co atoms would be approximately 0.84 using the cross-sections of 4.6 barns for ^{58}Ni (n, γ) and 37 barns for ^{59}Co (n, γ).

The ^{60}Co activity in 68 g sediments was counted to be $\leq 0.34 \times 10^{-2}$ cpm. Since double ^{60}Co half-lives had already passed since the dredged time (February, 1968), the ^{60}Co activity at the dredged time might have been $\leq 1.2 \times 10^{-2}$ cpm, corresponding to approximately $\leq 2.3 \times 10^7$ (^{60}Co atoms/kg sediments), using a counting efficiency of 3.0%. The number of neutron induced ^{59}Ni atoms is then less than $0.84 \times 2.3 \times 10^7$ (atoms/kg sediments). The specific activity of this neutron induced ^{59}Ni would be $\leq 3.4 \times 10^{-4}$ (dpm/kg sediments), which is less than 1% of the measured activity obtained in this work.

According to Furukawa (1980), the activity ratio [$^{60}Co/^{137}Cs$] was 0.08 in the fallout due to nuclear bomb explosions in the atmosphere and was approximately 2.0 in shallow sea sediments dredged from the vicinity of atomic power plants. In our sediment the upper value of ^{137}Cs atoms was measured as 0.2×10^7 (atoms/kg sediments). Using the extreme value of [$^{60}Co/^{137}Cs$] = 2.0, the ^{59}Ni activity expected from nuclear bomb explosions and atomic power plants would be less than 3.8×10^{-4} (dpm/kg sediments).

This value was calculated for the extreme case, but agrees by chance with the value mentioned above.

In any case, the contribution of the ^{59}Ni fraction due to artificial sources to the cosmic ray produced ^{59}Ni fraction is expected to be negligibly small.

^{53}Mn in deep-sea sediments

^{53}Mn (3.7×10^6y) of cosmic origin in deep-sea sediments was measured by Imamura *et al.* (1979) using an isotope separation process and neutron activation analysis. A value of (^{53}Mn/^{10}Be) = 0.0016 ± 0.0007 was obtained.

^{53}Mn is one of the most useful and long-lived radionuclides produced by solar cosmic rays, such as a reaction of ^{56}Fe (p, alpha), in cosmic dust and then accreted onto the Earth's surface. However, as described in Chapter 7, atmospheric heating causes a prompt evaporation of induced Mn atoms from the iron bodies, so that the hot atoms of Mn condense and fall down onto the Earth's surface.

The residence time of Mn atoms in sea water is not very long compared with the half-life of ^{53}Mn, which is estimated as about a few hundred years.

A highly sensitive detection technique for ^{53}Mn is the neutron activation analyzing technique: ^{53}Mn (n, gamma) ^{54}Mn (312.5 days, 835 keV) (Millard Jr. 1965).

This method is indeed very sensitive; however, it is necessary to pay attention to some side reactions, such as ^{54}Fe (n, p) ^{54}Mn and ^{55}Mn (n, 2n)^{54}Mn. The Mn content in deep sea sediments is so abundant that Imamura *et al.* (1979) tried to concentrate the mass = 53 component using a large scaled mass separator (Table 5.12).

Imamura *et al.* (1979) tried to calculate the global accretion rate of ^{53}Mn as $(0.7 \pm 0.3) \times 10^5$ (dpm/m^2, yr). The (^{53}Mn/^{10}Be) ratio and the accretion rate of the cosmic matter estimated from Ir and Os determinations in deep-sea sediments suggests the undersaturation of ^{53}Mn in cosmic dust materials, and also that the cosmic ray exposure age will be ~0.7 MA (million years). This age is in good agreement with those obtained in Chapter 6.

5.5 Cosmic Ray Exposure Age Determination by Measurements of ^{59}Ni and ^{53}Mn Intensities in Deep-Sea Sediments

^{59}Ni and ^{53}Mn intensities in deep-sea red clays, which were considered to be carried by small-sized extraterrestrial dust into the sediments, were determined (Yanagita, Yamakoshi and Imamura 1979). In this work, the ^{59}Ni was measured directly by a large area, soft X-ray detector, and the ^{53}Mn was measured by the enrichment method mentioned in Section 5.4 using a mass separator. After the mass separation, they were irradiated by neutron flux for a long time. After a radiochemical purification, they were counted by a low-level radiation counter.

A "mean" cosmic ray exposure age of cosmic dusts contained in sediments could be calculated by the following formula. In this case we do not know how much cosmic matter is contained in the sediments used. However, the total amount of cosmic matter is assumed to be D grams in the sediments.

Table 5.12. Typical results obtained by the enrichment-NAA method by Imamura *et al.* (1979) in two sedimental cores, which are already dated by the ^{10}Be method, are shown below.

SAMPLE DEPTH (cm)	ISOTOPE ENRICHMENT FACTOR	^{53}Mn (μgm)	{ENRICHED ^{53}Mn SAMPLES} cpm [measured]	^{54}Mn/mgm Mn [from ^{53}Mn]	^{53}Mn in Sediment [dpm ^{53}Mn] ($\times 10^{-5}$)	Cores (10^{-3} dpm/gm Mn)
SAM-N 20~28	324	100	0.165 ± 0.017	0.037 ± 0.018	3.2 ± 1.5	1.06 ± 0.51
SAM-S 0~40	416	112	0.145 ± 0.021	0.017 ±0.022	1.7 ± 2.1	0.38 ± 0.48

(1) The correction for ^{54}Fe (n, p) ^{54}Mn was negligible. (2) The background rate was 0.0172 ± 0.0007 cpm. (3) The counting efficiency was 4.7% for the photo peak of 834.8 ± 4 KeV. (4) The beam implantation efficiency was 0.93 ± 0.7%. (5) The correction was 0.128 ± 0.006 cpm (^{54}Mn/mg Mn) for the ^{55}Mn (n, 2n) ^{54}Mn reaction.

The atom numbers of a radionuclide produced by cosmic ray irradiation in outer space are expressed as:

$$dN/dt = g - \lambda N,$$

where N = the atom numbers of induced nuclides per unit mass of dust, g = the production rate of the nuclides in cosmic dust per unit time, λ = the mean life of the radionuclide. Integration of the formula gives:

$$N = (g / \lambda)[1 - \exp(-\lambda T)],$$

where T = the cosmic ray exposure age of the "mean" value for all sized dusts in the sediments. The assumed total weight of the dust contained in the measured sediments is D, so that the total atom numbers are ND, and the disintegration rate is λND (dpsec). If we have the intensity ratio [R] of two long-lived radionuclides (1) and (2) derived from the same iron matter:

$$R = \frac{\lambda_1 N_1 D}{\lambda_2 N_2 D} = g_1[1 - exp(-\lambda_1 T)] / g_2[1 - exp(-\lambda_2 T)],$$

where $g = n\sigma f$ (atoms/sec., g dust), n is the atom number of the target material, σ is the cross section for the production of the induced nuclide, and f is the flux of cosmic radiation.

For larger bodies, such as meteorites, the main contributions come from the processes of spallation and radiative capture.

The radiative process is especially dominant in stony meteorites. However, for smaller meteoroids, such as dust-particles or micrometeorites, the situation is much different. The contribution from the proton and alpha particle induced processes are much more dominant. The solar proton flux was used from the studies in lunar samples (Reedy and Arnold 1972). At a value of R = 100 MV for the characteristic rigidity, the solar proton flux is

$$J(> 10 MeV, 4\pi) = 100(protons / cm^2, \text{sec.}).$$

The ratio of proton to alpha particle flux used in this work is 22 ± 4 (van Hollenbecke 1975). According to Lal and Venkatavaradan (1967), the induced ^{59}Ni could be recalculated using the obtained cross-section (Yanagita *et al.* 1978) and the p/alpha ratio of 0.29 (dpm/g, dust). In the above calculations it is assumed that the ^{59}Ni and ^{53}Mn activities in cosmic dust were in secular equilibrium for a long time at 1 AU. However, this assumption is incorrect, as will be shown soon. And we have also assumed 25% of iron, 1.35% of Ni and 520 ppm of Co as the chemical composition of the cosmic dust.

The dredged sample used was collected in the Central Pacific Ocean. After a large-scale chemical extraction, the obtained nickel fraction was electroplated onto cylindrical holders of a proportional counter whose area was 312 cm^2. The amount of the nickel fraction was 430 mg, so that the thickness was 1.38 (Ni mg/cm^2).

The extracted manganese fraction was concentrated beforehand with a mass separator in the chemical form of $MnCl_2$. A manganese sample obtained at $(m/e) = 53$ position in a mass separation process and also blank samples of ^{55}Mn for (n, 2n) and ^{54}Fe for (n, p) reactions were investigated in a high flux thermal neutron reactor for three months or more. After purification, each sample was counted by a low level gamma ray counter system.

^{59}Ni activity was obtained as $(5.53 \pm 1.66) \times 10^{-2}$ (dpm/kg sediments). ^{53}Mn samples were determined as (1); $(6.9 \pm 3.1) \times 10^{-3}$ and (2); $(3.8 \pm 2.7) \times 10^{-3}$ (dpm/kg sediments). The weighted mean value of runs (1) and (2) was $(5.1 \pm 2.0) \times 10^{-3}$ (dpm/kg sediments).

So we could obtain the activity ratio R, and we could calculate the "common" cosmic ray exposure age T to be 3.2×10^5 yr.

In this calculation of the cosmic ray exposure age, we used the cosmic ray intensities at 1 AU. However, the ratio value (g_1/g_2) is fairly independent of the spatial distribution of solar cosmic rays. The ratio value (g_1/g_2) is written in detail as $(n_1\sigma_1 f_1/n_2\sigma_2 f_2)$, we know n and σ. From the (f_1/f_2) ratio, the history of the dust motion or trajectories which occurred before accretion onto the Earth is cancelled out.

Compared to the half-life of ^{53}Mn (3.7×10^6 yr), the "mean" exposure age was so much shorter, on carriers which were meteoroids in interplanetary space, that the ^{53}Mn activities must be in undersaturation.

Supposing two size-distribution patterns of the dust, Venkatavaradan (1970) calculated the saturation factors of long-lived nuclides in cosmic dust, starting from 3 and 10 AU, moving under the Poynting-Robertson effect in space.

In this work the saturation factors were estimated to be 0.058 for ^{53}Mn and 0.95 for ^{59}Ni. These values did not agree with those by Venkatavaradan (1970); however, the half-life of ^{53}Mn used was much shorter in his work. The cosmic ray exposure ages of the dust are discussed in Chapter 6 as well.

REFERENCES

Alvalez W., Alvalez L. W., Asaro F. and Michel H. V. 1980 Science 208 1095.
Amano S., Okada A. and Shima M. 1967 Bull National Mus. Tokyo 10 471.
Amin B. S., Kharkar D. P. and Lal D. 1966 Deep-Sea Res. 13 805.
Amari S. and Ojima M, 1985 Nature 317 520.
Angino E. E. 1966 Geochim. Cosmochim. Acta 30 939.
Barker Jr. J. R. and Anders E. 1968 Geochim, Cosmochim. Acta 32 627.
Dostrovsky I., Frenkel Z. and Friedlander G. 1959 Phys. Rev. 116 683.
Fujita Y. Taguchi Y., Imamura M. Inoue T. and Tanaka S. 1975 Nucl. Instr. Meth. 128 523.

Fukumoto H., Nagao K. and Matsuda J. 1986 Geochim. Cosmochim. Acta 50 2245.
Goldberg E. D. 1954 J. Geology 62 249.
Goldberg E. D. and Arrhenius G. O, S. 1958 13 153.
Goldberg E. D. 1963 The Oceans as a Chemical System, The Sea, ed. M. N. Hilll (Inter-Science Publish.)
Harriss R. C, Crocket J. H. and Stainton M. S. 1968 Geochim. Cosmochim. Acta 32 1049.
Hodge P. W. and Wright F. W. 1964 J. Geophys. Res. 69 2449.
Hunter W. and Parkin D. W. 1959 Proc. Roy. Soc. 255A 382.
Imamura M., Inoue T., Nishiizumi K. and Tanaka S. 1979 Proc. Intern. Conf. Cosmic Ray (Kyoto) OG-12-21.
Inoue T. and Tanaka S, 1976 Earth Planet, Sci. Lett. 29 155.
Inoue T.and Tanaka S. 1978 Nature 277 209.
Inoue T. Tanaka S. 1979 Nature 277 209.
Inoue T. and Tanaka S. 1982 Geochem. J. 16 321.
Kobayashi K., Kitazawa K., Kanaya T. and Sakai T. 1971 Deep-sea Res. 18 1045.
Laevastu T. and Mellis 0. 1955 American Geophy. Union 36 385.
Laevastu T. and Mellis 0. 1961 J. Geophys. Res. 66 2507.
Lal D. and Venkatavaradan V. S. 1967 Earth Planet. Sci. Lett. 3 299.
McCorkell R., Fireman E. L. and Langway Jr. C. C. 1967 Science 158 1690.
Merrihue C. 1964 Ann. New York Acad. Sci. 119 351.
Murray J. and Renard A. F. 1891 Extraterrestrial Minerals, Challenger Report Chap. 4.
Nogami K., Yamakoshi K. and Ninagawa K. 1976 Nucl. Inst. Meth. 152 195.
Oepik E. J. 1955 Nature 177 929.
Ojima M, Takayanagi M., Zashu S. and Amari S 1984 Nature 311 449.
Otsuka H., Tazawa Y., Yamakoshi K. and Yasuda S. 1972 INSJ137 (Univ. Tokyo).
Parkin D. W. and Tilles D. 1968 Science 159 936.
Pettersson H. and Rotschi H. 1950 Nature 166 308.
Pettersson H. and Rotschi H. 1952 Geochim. Cosmochim. Acta 2 81.
Pettersson H. 1959 Geochim. Cosmochim. Acta 17 209.
Raisbeck G. M. and Yiou F. 1981 Nature 292 825.
Rajan R. S., Browlee D. E., Tomandl P., Hodge W., Farrar A and Britten R. A. 1977 Nature 267 133.
Reedy R. C. and Arnold J. R. 1972 J. Geophys. Res. 77 537.
Reyss J. L., Yokoyama Y. and Tanaka S. 1976 Science 193 1119.
Smales A. A., and Wiseman J. D. H. 1955 Nature 175 464.
Smales A. A., Mapper D. and Wood A. J. 1957 Analyst 82 75.
Sverdrup H. U., Johnson M. W. and Fleming R. H. 1942 The Oceans, Their Physics, Chemistry and General Biology (Prentice Hall Inc.)
Takayanagi M. and Ojima M. 1987 J. Geophys. Res. 92 B12 12531.
Tanaka S., Sakamoto K., Takagi J. and Tsuchimoto M. 1968 Science 16 0 1348.
Tanaka S., Inoue T. and Imamura M. 1977 Earth Planet. Sci. Lett. 37 55.
Tatsumoto M. 1956 Nihon Kagaku Zasshi 77 1637 (in Japanese)
Turekian K. K., 1958 Nature 182 1728.
van Hollenbecke M. A. 1975 Proc. Intern. Cosmic Ray Conf. 5 1563.
Venkatavaradan U.S. 1970 Thesis of Univ. Bombay (unpublished).
Wahlen M. 1969 Thesis of Bern Univ. (unpublished)
Wasson J. T. 1963 ICARUS 2 54.
Yamakoshi K. 1971 Memoirs Faculty Sci., Kyoto Univ. 33 311.
Yamakoshi K. and Tazawa Y. 1971 Nature 233 542.
Yamakoshi K. and Nogami K. 1976 Nucl. Instr. Meth. 134 519.
Yamakoshi K. 1988 Abstracts 19th LPSC (Houston) 1306.

Yamakoshi K. and Yanagita S. 1981 Earth Planet. Sci. Lett. 52 259.
Yanagita S., Yamakoshi K. and Gensho R. 1978 Nucl. Phys. A303 354.
Yiou F. and Raisbeck G. M. 1972 Phys. Rev. Lett. 29 372.

Chapter 6

Studies on Spherules Collected from Ice Layers and Deep-Sea Sediments

6.1 Research History

In 1874 Nordenskioeld, a Swedish geologist and explorer, found a small quantity of dark magnetic powder in snow samples which was proven to contain metallic iron, cobalt and probably nickel. He expressed his belief that the magnetic material was of cosmic origin.

In 1876 Murray deduced that magnetic spherules collected from deep-sea sediments taken during the Challenger Expedition, from 1874 to 1876, were of meteoric origin. He called them "cosmic spherules"; however, the origin of the samples was not confirmed and extended until the last twenty years.

From deep sea sediments, Pettersson and Rotschi (1950) and Laevastu and Mellis (1955) separated magnetically metallic spherules and found that they were similar to those described by Murray. Smales *et al.* (1957), using a neutron activation method, found that the nickel-cobalt ratio in the collected spherules was also similar to that in iron meteorites.

Merrihue (1984) found rare gas anomalies in magnetic fractions collected from deep-sea sediments. Fechtig and Utech (1964) filed up the spherules and obtained cross-sections of the samples. They measured the nickel contents in the nuclei of the spherules and found that spherules of the Tertiary Period had no nuclei and no nickel content.

Millard and Finkelman (1970) found siderophile elements using instrumental neutron activation analysis (INAA), and Fe, Ni, Co and traces of Ir were detected.

After their work, detection of Ir contents was considered as good evidence for the samples being of extraterrestrial origin.

Up to now, various fundamental analyzing methods and techniques, which have already been successfully applied to studies on meteorites and lunar samples, have been tested and applied to spherule researches; cosmogenic radionuclide detection (Raisbeck *et al.* 1983), mass analysis of isotopic anomalies (Shimamura *et al.* 1977, Shimamura *et al.* 1979), ablation experiments in laboratory simulations (Nogami 1985, Nogami and Yamakoshi 1985) and so on. These problems and the answers are described in the following sections, and we could obtain their cosmic ray exposure ages and deduced their origins in the interplanetary space (Yamakoshi, 1991).

After the enthusiastic research on the Apollo and Luna projects, the precursor and inhomogeneity problems of the primordial solar system have been raised as the most interesting exercises in cosmochemistry. In the 1970's, Japanese cosmophysicist groups succeeded in establishing detection methods of chemical and isotopic compositions in microfine spherules using various scientific methods (Nogami *et al.* 1980, Yamakoshi *et al.* 1981).

Yamakoshi *et al.* (1981) measured systematically over 400 metallic spherules and examined the relation between size and chemical composition of Ir, Co and Au. In the work, they analyzed using INAA 100 iron spherules whose sizes were larger than 100 μm and concluded that 85 particles were confirmed to be of extraterrestrial origin. In the following studies (Yamakoshi 1985) he measured a large number of iron spherules, whose sizes ranged from 16 to 1440 μm [see section 6.3]. And a suggestive conclusion is obtained from the results.

REE abundance patterns in glassy spherules were measured, and similar patterns just like those of primitive chondrites were found by Nagasawa *et al.* (1979). Chondritic, stony spherules were analyzed with INAA by Yamakoshi (1984) and deduced as being of cosmic origin, because of the great abundance of noble metals in them.

6.2 Size and Elemental Composition of Cosmic Spherules

Magnetic iron spherules were sampled from deep-sea sediments. The major part of the spherules is thought to be of extraterrestrial origin, because of their chemical composition as well as the cosmogenic radio- and stable nuclides contained in the spherules. These spherules are expected to be able to offer a missing connecting-link between information about extremely small stratospheric dust grains and micro-meteorites, whose sizes are larger than spherules.

However, the cosmic matter of these sizes usually suffers the most thermal degeneration during atmospheric entry. Larger bodies, such as meteorites, change composition in just the thin surface layers, and smaller ones, such as stratospheric, fluffy grains, can be cooled and maintain body temperatures under the melting point because they have a very large area relative to their mass. Therefore, much appropriate laboratory simulation experiments are necessary to learn the behavior of refractory and volatile elements of the pre-atmospheric compositions of molten cosmic bodies. The chemical compositions of iron spherules are shown in Table 6.1.

Table 6.1. Chemical compositions of five iron spherules gathered from deep sea sediments. (Yamakoshi 1983).

SAMPLE	SIZE (μm)	WEIGHT (μgm)	Fe (%)	Ni (%)	Co* (%)	Au (ppm)	Ir* (ppm)	Os (ppm)	Ru (ppm)
#23	530	293	70.1 ± 0.2	10.8 ± 0.6	0.062	0.22 ± 0.011	5.7	23.9 ± 0.5	2.6
#24	500	303	67.1 ± 0.1	5.6 ± 0.2	0.26	0.083 ± 0.018	3.0	9.4 ± 0.2	5.8
#28	500	255	82.9 ± 0.2	1.6 ± 0.1	0.13	< 0.003	7.0	20.4 ± 0.2	10.9
#29	540	401	83.7 ± 0.3	0.5 ± 0.1	0.11	< 0.01	7.0	20.7 ± 0.1	4.7
#30	440	250	75.8 ± 0.3	2.7 ± 0.2	0.47	0.03 ± 0.01	3.1	9.7 ± 0.4	8.9

*In the cases of Co, Ir and Ru the statistical errors can be neglected.

The elemental compositions vs. spherule sizes have been determined by INAA (Yamakoshi 1985). The spherules whose sizes ranged from 16 μm to 1440 μm were gathered from red clay. The ratios of (Co/Fe), (Ir/Fe) and (Au/Fe) were obtained from the data. The deep-sea sediments used in this work were dredged by R/V Hakurei-Maru off the Hawaiian Islands at more than 4500 meters in 1979. We picked out the spherules from the magnetic fractions gathered from the red clay sediments by a hand-magnet. There is no bias for selection of the spherules from the magnetic fraction. However, in the smallest size region we could hardly pick up the spherules, because some of the spherules were so brittle that a tweezer sometimes crushed them. Therefore, in the smallest size region hard and solid spherules were liable to be selected.

The siderophile elemental concentrations of Fe, Co, Ir and Au were determined by INAA. In the determinations, ^{59}Fe (half-life = 44.6 d), ^{60}Co (5.271 y), ^{192}Ir (74.02 d) and ^{198}Au (2.696 d) activities were determined by a low-background, large-volume Ge (Li) detector equipped with a multi-channel pulse height analyzer.

Neutron irradiations were performed on larger spherules for a few hours in a TRIGA II reactor, whose flux was ca. 0.7×10^{11} (n/cm^2, sec). For smaller ones the irradiations were done for 11 days in a research reactor whose flux was ca. 3×10^{13} (n/cm^2, sec).

Reference samples were made of an iron meteorite, Canyon Diablo; the homogeneity of its elemental composition had been already examined. The obtained data are shown in Figs. 6.1~6.3.

The experimental errors were estimated as follows. The obtained counts from ^{59}Fe and ^{60}Co in the almost all spherules, even those smaller than 100 μm size, were enough to achieve ±5% statistical error (±σ). The errors of ^{192}Ir and ^{198}Au, however, could not be definitely determined. For errors poorer than ±100%, the points marked by arrows are given as the upper limits shown in the figures.

The accumulated uncertainty caused by the reproducibilities of the radiation countings due to the sample mountings on a detector and by inhomogeneity of the reference samples did not exceed ±10% for any point in the figures.

The data obtained by our research group are plotted in the figures. Co, Ir and Au concentrations were normalized to the Fe concentrations, because we could eliminate the amounts of oxygen in oxidized samples and also because we could not weigh an individual spherule weight (<0.2 μg) with a balance.

The origin of the deep-sea spherules is not decisively determined; they may be ablated grains of micrometeorites, the molten droplets from meteorites due to atmospheric heating or even terrestrial volcanic debris or artificial contaminants. In all cases they must have been heated up to the melting point once. However, the spherules with high contents of siderophile elements can be considered to be of extraterrestrial origin. The metal phases in chondrites and iron meteorites are available candidates for source materials of extraterrestrial iron spherules.

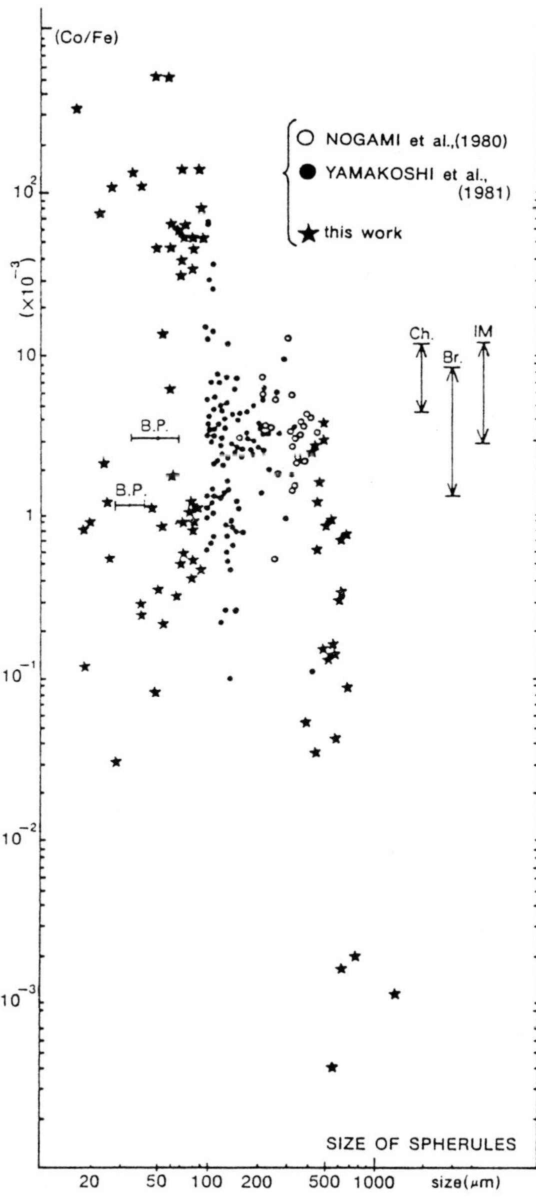

Fig. 6.1. Spherule size vs. (Co/Fe) of iron spherules (Yamakoshi, 1985). B.P. show the data from Ganapathy *et al.* (1978).

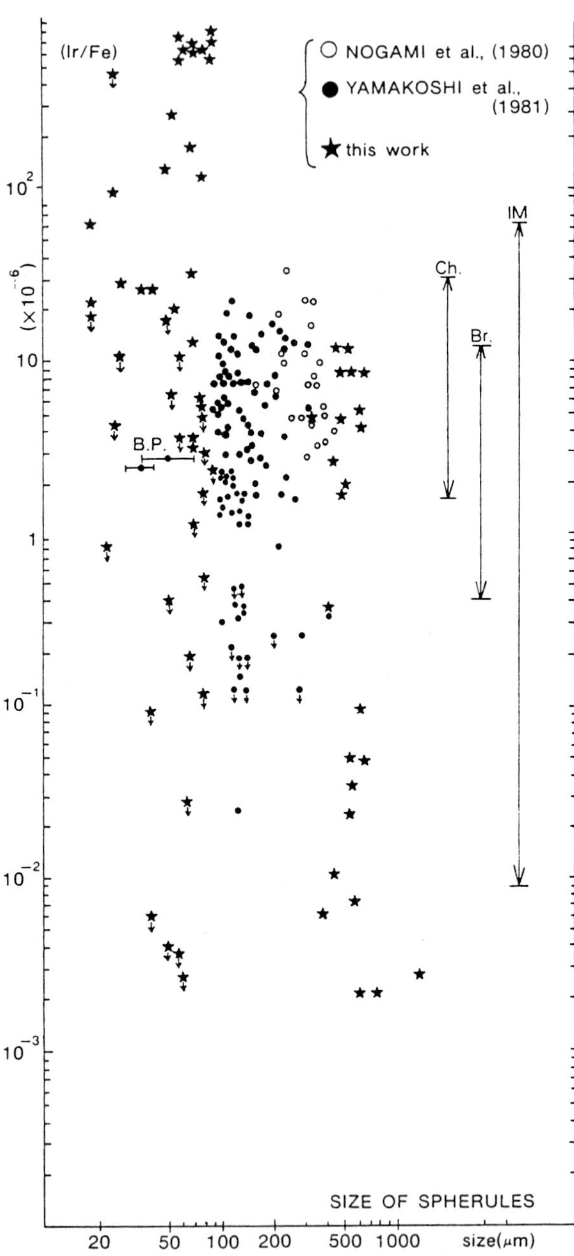

Fig. 6.2. Spherule size vs. (Ir/Fe) in iron spherules (Yamakoshi, 1985).

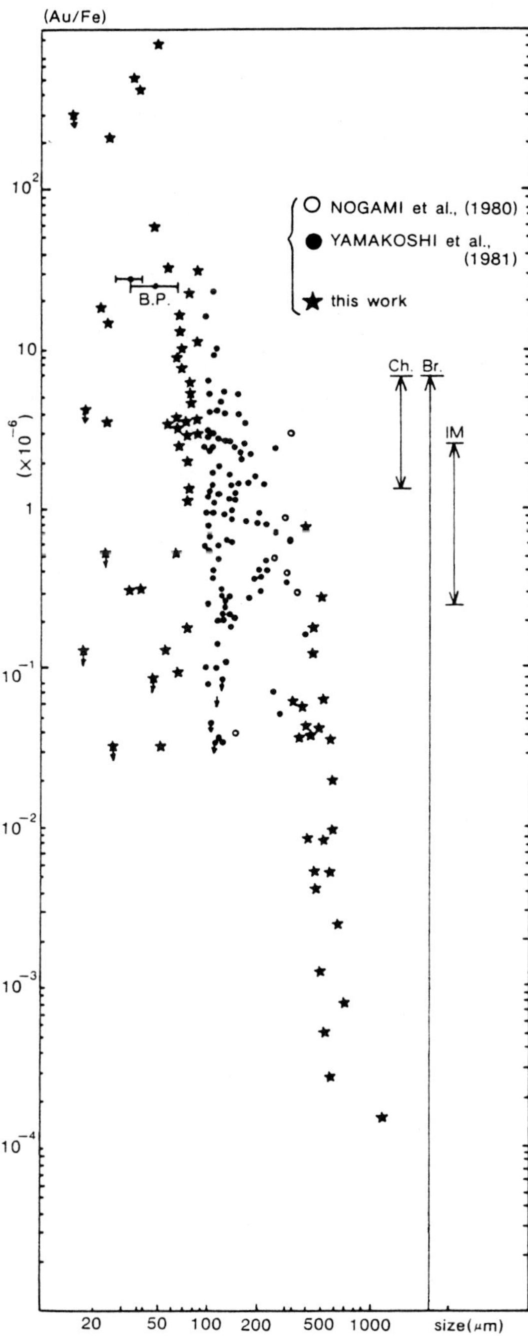

Fig. 6.3. Spherule size vs. (Au/Fe) in iron spherules (Yamakoshi, 1985).

In order to assess the possible origin of the spherules sampled from deep-sea sediments and ice layers in polar regions, we determined the siderophile element contents, such as Co, Ni, Ir and Au, of individual spherules and compared these values with the available meteoritic and terrestrial crustal rock values. The comparison of meteorites to crustal igneous rocks for each element shows that siderophile elements are much more abundant in chondritic meteorites and were even enriched in most iron meteorites. This is derived from the fact that the Earth is fractionated and altered chemically and thermally. Chemical and thermal fractionation do not occur so much or so completely in the parent bodies in space. Therefore, the siderophile elemental compositions of meteoroids are relatively higher than those of the Earth's crust.

In a previous work (Yamakoshi *et al.* 1981) we gathered 735 iron spherules from deep-sea sediments of 1.83 kg in dry weight. One hundred spherules of larger than 100 μm in size were analyzed with INAA; Fig. 6.4 shows a histogram in 20-μm intervals of 735 iron spherules. There are relatively large uncertainties in the counts of the smaller spherules (≤40 μm) because of difficulties in picking them out from the magnetic fractions. As shown in the figure, except at the smallest sizes the inverse third power law is acceptable for describing the size distribution of the spherules (Fig. 6.4).

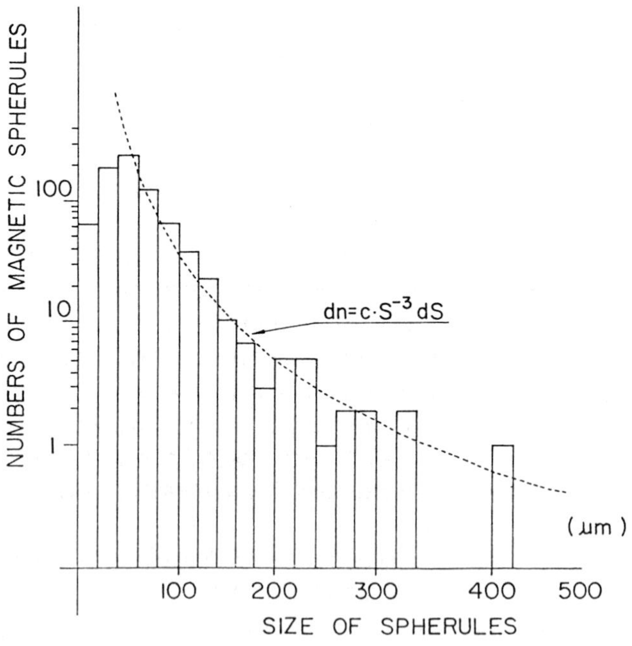

Fig. 6.4. Size interval of 735 iron spherules; 100 spherules of larger than 100 μm in size were examined; 85 spherules in tested 100 cases were confirmed to be of extraterrestrial origin (Yamakoshi *et al.* 1981).

$dn \propto S^{-P}\, dS = S^{-3}\, dS$, where dn is the number of spherules with sizes in the range S to $S + dS$. The error in the exponent P is estimated from the uncertainty in the gradient determination in Fig. 6.4; we estimated $P = 3 \pm 0.2$.

The number of the spherules larger than 80 μm was 168. The lower limit, 80 μm, is taken as the smallest size for which there is negligible bias in the picking-out procedures. The largest spherule was 440 μm in size.
Therefore

$$N = \int dn = C\int_{0.008}^{0.044} S^{-3} dS = 168,$$

and it follows that the normalization constant is $c = 2.22 \times 10^{-2}$.

Assuming a common density ρ of 5.2 for all iron spherules (in iron oxides; the densities are 5.2 for magnetite, 5.3 for hematite and 5.2 for maghemite), the total weight of the spherules in 1.83 kg of dry sediments should be contained by the following integral calculation:

$$W = \int_0^{0.044} dn\left(\frac{4}{3}\right)\pi\left(\frac{S}{2}\right)^3 \rho = \frac{\pi}{6}\rho C\int_0^{0.044} dS = 2501(\mu g).$$

Therefore, the concentration of iron spherules in dry red clay sediment is

$$\left(2.50\times10^{-3} / 1.83\times10^{3}\right) = 1.36\times10^{-6}\left(g / g \text{ of dry sediments}\right)$$

If we assume a constant accretion rate of meteoroids onto the Earth, the concentration of the spherules in sediments depends on the sedimentation rate. In this case the sedimentation rate is unknown; however, the value of the concentration (1.36 mg/Kg of sediments) obtained here coincide with the results (1.1 mg/kg of sediments) determined by Laevastu and Mellis well (1955), whose sedimentation rate was estimated as 0.4 mm per 1000 years.

It is possible to make some comparisons with previous works. Laevastu and Mellis (1955) reported finding no spherules larger than 230 μm, which was in agreement by chance with theoretical considerations by Oepik (1955).

According to Oepik's calculations, based on the collision theory of cosmic bodies and the Poynting-Robertson Effect, essentially all cosmic particles larger than 250 μm in diameter are intercepted by the planet Jupiter and therefore they do not reach the Earth. However, lunar microcraters and meteor studies have shown that meteoroids larger than 250 μm do exist at 1 AU, and also we have discovered spherules in some studies which exceeded Oepik's cutoff. Of course, it is possible that larger dust grains are supplied only from sources other than those considered by Oepik (fragmentations of asteroids or cometary supplies) however, Fig. 6.4 does not show any drastic changes of the elemental ratios with size.

With regard to the size distribution, Laevastu and Mellis (1961) obtained an empirical formula for iron spherules from deep-sea sediments.; with V, the volume of a single spherule in a given size group and N, the number of spherules in this size group, they obtained the relation $V \times N$ = constant, which is in accord with our results (Yamakoshi *et al.* 1981). Fireman and Kistner (1961) also applied an inverse third power distribution ($dn \propto S^{-3}\, dS$) to studies on magnetic spherules sampled with airplane-borne dust collectors.

However, Brownlow *et al.* (1966) obtained a size distribution for iron spherules ranging between 35 and 120 μm in size, which in terms of mass was given as

$$dn \propto m^{-1.53} dm.$$

A mass distribution $dn \propto m^{-5/3}\, dm$ can be readily obtained from the size distribution $dn \propto S^{-3}\, dS$, assuming a common density for all rounded spherules; thus with this assumption the work of Brownlow *et al.* (1966) implies a different distribution; $dn \propto S^{-2.59}$. Assuming a common density for the spherules, our results imply a differential distribution, $dn \propto m^{-5/3}$ dm.

Parkin and Tilles (1968) summarized the results on the size and mass distributions from zodiacal light observations, cometary meteor countings and theoretical consideration. Dohnanyi (1978) reviewed a differential mass distributions of incoming objects into the Earth's atmosphere, and exponents of –13/6 (for cosmic bodies heavier than 0.1 μg) and –1.5 (lighter than 0.1 μg) were proposed.

Table 6.2. The mass exponents of the size distribution curves (Brownlow *et al.* 1966).

DEPTH (cm)	TOTAL SPHERULES A, 30~99 μm	TOTAL SPHERULES B, >100 μm	RATIO A/B	DIFFERENTIAL MASS EXPONENT
0~90	43	4	10.7	–1.65
100~190	60	18	3.3	–1.33
200~290	86	10	8.6	–1.60
300~370	66	23	2.9	–1.29
380~430	79	17	4.6	–1.42
440~490	79	10	7.9	–1.57
500~570	81	12	6.7	–1.53
568~650	107	14	7.7	–1.56
660~710	90	10	9.0	–1.60
720~790	131	10	13.1	–1.71

There are two difficulties in comparing the spherule distribution directly with the data obtained from the zodiacal light and meteor studies: [1] the spherules are not confirmed completely as of extraterrestrial origin, and [2] the volume, size and shape of cosmic dust are changed by thermal degeneration through frictional heating in the Earth's atmosphere.

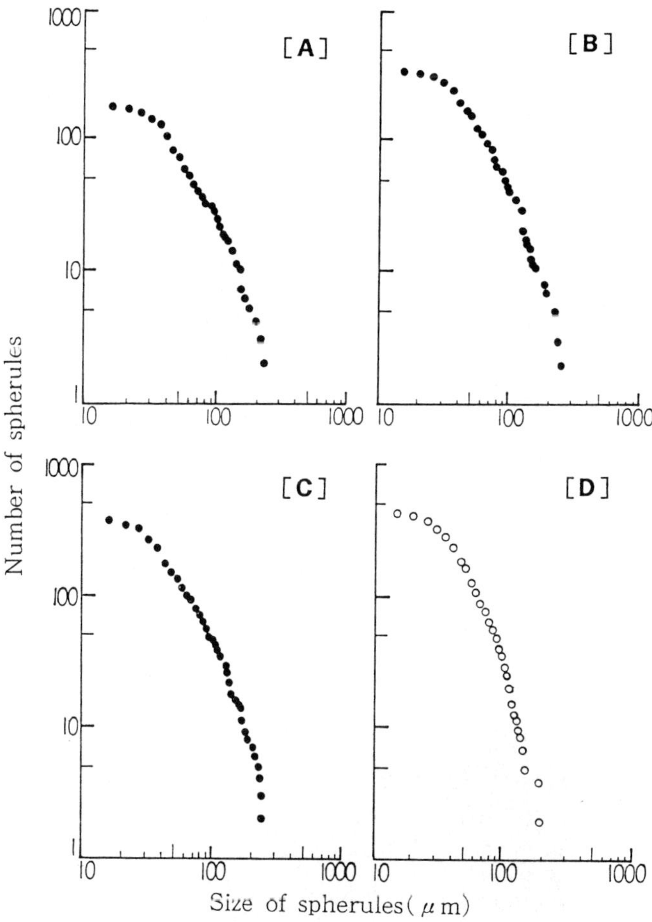

Fig. 6.5. Size integral distribution for all spherules in 2 meter portions down the core; A[0~190 cm], B[200~390 cm], C[400~590 cm] and D[600~790 cm]. The ordinates represent the integral number of spherules larger than the grain size (Brownlow *et al.* 1966).

While Brownlow *et al.* (1966) corrected the size distribution of iron spherules in deep-sea sediments to the pre-atmospheric forms through thermodynamical calculations, we are pursuing thermal degeneration studies to obtain empirical size corrections. At the present time, however, a comparison of our results with those of Dohnanyi is premature.

Brownlow *et al.* (1966) gave us suggestive results on the size distribution of iron spherules. They separated a Pacific sediment core of 8.5 m in length into four portions: 0~190 cm, 200~390 cm, 400~590 cm and 600~790 cm. They picked out the metallic-centered and hollow magnetic spherules with sizes ranging 35 to 120 μm and tried to compare the size distribution of the four portions of the core mentioned above, which are shown in Fig. 6.5.

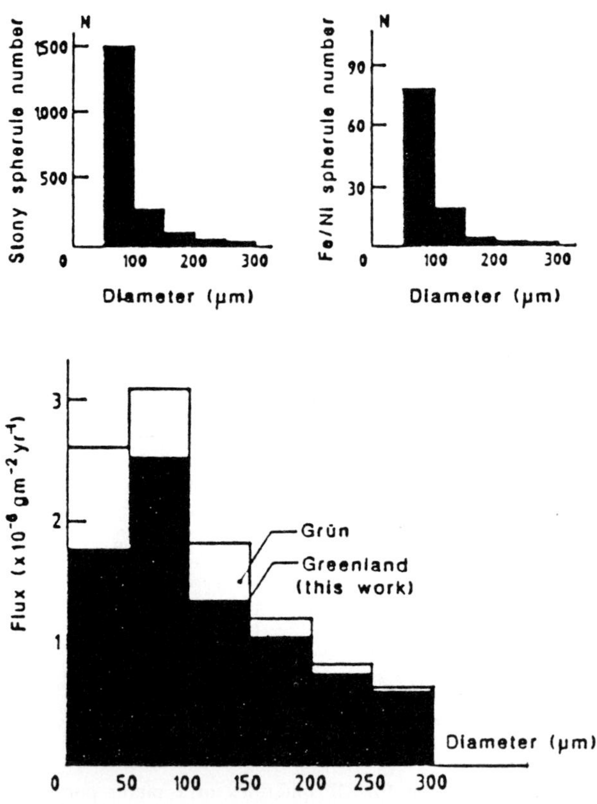

Fig. 6.6. Mass histogram of Greenland samples. The two top histograms give the size distributions of chondritic-stony spherules (left) and also oxidized Fe/Ni spherules (right). The lower histograms show the micrometeorite flux by Gruen *et al.* (solid line-white region) detected at 1 AU and the flux of the Greenland spherules (shaded region).

In these four portions, the slopes of cumulative curves in (A), (B), (C) are similar to each other. However, (D) is quite different; that is, the ratios for the number of spherules with size groups of (30 to 100 μm) and (≥100 μm) in A, B, C and D are 4.8, 5.4, 5.0 and 10.2, respectively. A core, sampled fairly close to this core sampling site, has been determined to have a sedimentation rate of ca. 1.5 mm/1000 years. Therefore, we can ask a question, a long time ago (more than 4×10^6 years before present) what happened? Why was the slope of the period so different from recent ones?

Mourette *et al.* (1987) examined the mass distribution of extraterrestrial dust collected from the Greenland ice cap. They stated that the grain mass distribution is very similar to that of micrometeorites at 1 AU, obtained by E. Gruen *et al.* (1985). This suggests that most of the grains are themselves micrometeorites and not ablation products of much larger meteorites.

In all grains Ni and high Ir concentrations could be measured, so they said, although no size formula was given (Fig. 6.6).

Nogami *et al.* (1988) gathered more than 1600 spherules from magnetic fractions in a water-tank sediment at the Australian Mawson Base in Antarctica. The water was piped from a pond behind the base area and supplied for drinking water. The sediment was supplied by the courtesy of Prof. R. Jacklyn of the University of Tasmania.

The size distributions were measured by an image processor device connected to a microscope.

Table 6.3. The sediment consisted of three samples (Nogami *et al.* 1988).

Sample code	Weight (wet) [kg]	Weight (dry) [g]	Magnetic fraction	Spherules	Spherules /g sediment
(1) clay	1.6	240	2.1 gm	785	3.3
(2) cobbles	2.0	130	1.8 gm	349	2.7
(3) clay	0.9	110	0.4 gm	477	4.3

The size histogram of the total iron spherules is shown in Fig. 6.7. The differential distribution is fitted to an experimental formula:

$$dn \propto S^{-2.3}\, dS,$$

which is rather different from that obtained from deep-sea spherules. The reason may be that artificial spheres exist in larger sizes.

The mass and size distribution of chondritic spherules from pelagic sediments were studied by Murrell *et al.* (1980). They collected 22 mg of stony

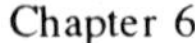

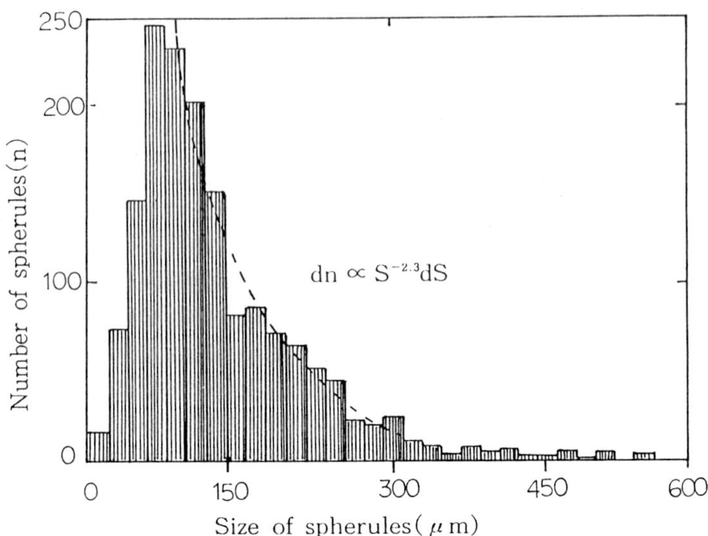

Fig. 6.7. The size histogram of iron spherules obtained from the sediment from the water tank of Australian Antarctic Base (Mawson).

and 50 mg of iron spherules larger than 150 μm from 411 kg (dry) of Pacific deep-sea sediments. It is uncertain whether all of them used in the work were of extraterrestrial origin or not.

They clustered in the mass range of 20~300 μg as the mass formula:

$$N\{> M(mg)\} = 5.13 \times M^{-1.88}.$$

In their statistics, the content of stony spherules larger than 200 μm is ~20 ppb in weight. The mass distribution ranged from 1 to 430 μg, and the mean mass was 24.5 μg. The size distribution ranged from 149 to 750 μm, and the mean size was 235 mm, because they used 200 mesh sieves for sediment selection. The shapes of the samples were divided into three categories: spheroidal (50%), ellipsoidal (~30%) and spherical.

The mean composition of 4.48 mg of a composite sample was determined by the atomic absorption method as [Mn] = 0.21%, [Fe] = 27%, [Co] = 0.005% and [Ni] = 0.62%. These data coincided surprisingly with those obtained by Yamakoshi (1985). They stated that the ratio of [Fe/Mn] = 130 is very similar to that of H chondrites.

The density histogram is shown in Fig. 6.8. The mean density is 3.12 g/cm^3.

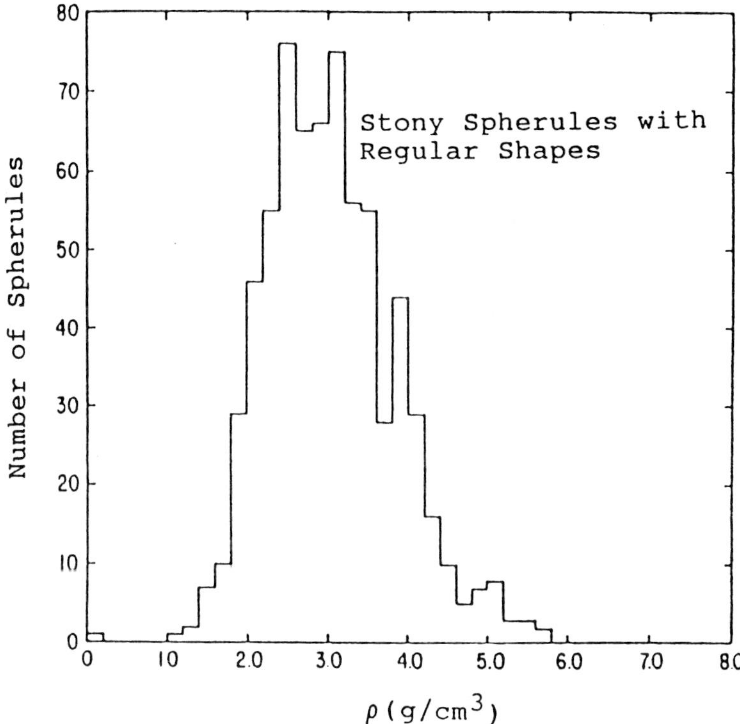

Fig. 6.8. Density histogram of the stony spherules regular in shape (Murrell *et al.* 1980)

Morphological research into the mass, size and/or density distribution of the collected spherules has been carried out in many studies. It is a well and widely known general theory on the size distribution of the "crushed" fragments of rocks and solid and homogenious materials that they distribute according to the inverse third power formula:

$$dn \propto S^{-3}\, dS,$$

where dn is the number of the crushed fragments between a size interval $S \sim S + dS$.

It is possible that the size distribution of pre-atmospheric dust takes the inverse third power form, since the origin of the dust might be from collision events with much larger bodies in space. There is no decisive proof that the size distribution of the residual debris of meteors, after flashing, spreads in the inverse third power. However, when abnormal changes of the power values of the distributions can occur, it is very interesting and suggestive that the ablation mechanisms occurred in the upper atmosphere.

6.3 Magnetic Iron Spherules

For the purpose of deciding the extraterrestrial origin of the magnetic iron spherules found in deep-sea sediments and ice layers of polar regions, the siderophile elements, such as Co, Ni, Ir and/or Au, were measured by instrumental neutron activation analysis (INAA). In general, siderophile elements such as Co Ni, Ir and Au, are much more abundant in extraterrestrial materials. Therefore, the detection of as much of these siderophile elements in a specimen as in meteorites is one of the most effective clues for determining their origins. It was a problem whether these spherules are ablated droplets removed from iron meteorites entering the Earth's atmosphere or cosmic iron grains themselves in space; however, the level of cosmogenic radioactivity detected is as large as estimated values in cosmic dust, which is much larger than those of larger iron meteorites. And X-ray diffraction analysis suggests that these spherules are the product of materials cooled rapidly.

For the small iron spherules of about 100 μg in weight, major elements, such as Mn, Fe, Co and Ni, have been determined with electron microprobe analysis by Larson *et al.* (1964) and by Schmidt and Keil (1966). They expressed the opinion that these spherules were of cosmic origin.

Neutron activation analyzing techniques have been applied to chemical composition determinations of iron spherules by Smales *et al.* (1957), Cassidy (1964) and Millard and Finkelman (1970). The INAA of iron spherules can be carried out in a non-destructive form; therefore, further investigations such as extremely highly sensitive mass spectrometry is applicable for the same spherules, which were done by Shimamura *et al.* (1977) and Shimamura *et al.* (1979).

In the work performed by Nogami *et al.* (1980), Fe, Co, Ni, Ir and Au were analyzed with INAA for individual spherules by low level gamma ray spectrometer systems. The isotopic ratios of ^{192}Ir to ^{194}Ir and the isotopic analysis of ^{185}Os/^{191}Os and ^{193}Os/^{191}Os were also investigated (Yamakoshi and Nogami 1991).

X-ray diffraction analysis for mineralogical investigations of the small iron spherules was done by Marvin and Einaudi (1967) and Millard and Finkelman (1970).

The most common mineralogy of iron spherules is an association of taenite [Ni-Fe] or native iron, wuestite [FeO], magnetite {Fe_3O_4] and hematite [Fe_2O_3] (Blanchard *et al.* 1980), and in many cases nuclear cores (Ni and noble metals enriched) and shell-crust structures (magnetite) can be observed.

The magnetic, iron spherules are obtained from magnetic fractions by picking out them with fine tweezers under a microscope. The magnetic fractions are gathered from the deep-sea sediments or ice layers of the polar regions with a hand-magnet. In this work, a large number of iron spherules with diameters ranging from 20 to 1440 microns were picked out. We classified the collected spherules roughly into the following three types:[1] spherules with rough surfaces very brittle to tweezer pressure. The specific gravities of the spherules are fairly low; [2] spherules with smooth surfaces, dull black in colour. These spherules have magnetic properties; however, the iron contents are so low that they are called as "chondritic"; [3] shiny spherules that are smooth enough to reflect the image of the tweezers. They are metallic gray or shiny black in colour. We have found that there are occasionally small craters or cavities on the surfaces of the spherules. The samples are weighed by an electrobalance. The specific gravities of spherules are calculated by a gravimetry method and also Stoke's method.

Figure 6.9 shows the typical features of spherules examined by a scanning electron microscope (SEM). Spherules are stuck on an aluminum disc with silver conducting paste for SEM samples no coating is needed on the spherules because they are already sufficiently conductive. Sometimes holes which look like craters or ripple patterns are seen on the surface of the spherules.

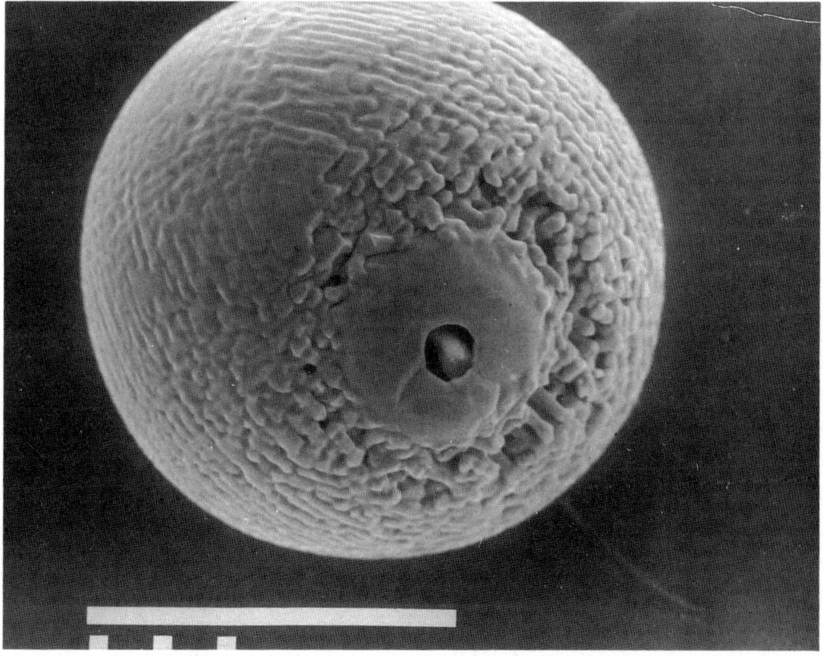

Fig. 6.9. SEM picture of iron spherule by courtesy of Prof. K. Nogami. (The scale bar represents 500 μm)

It is important that all desired elements are determined simultaneously in the same samples nondestructively to conserve materials for another analysis. Detectable elements in iron spherules by the INAA method within ±5% of the $\pm 1\sigma$ statistical errors are Fe, Co, Ni, Ir and Au with the sensitivity in the following ranges: Fe; 1 μg, Co; 1 ng, Ni; 5 ng, Ir and Au; 5 pg and other platinum groups, such as Pt, Os, Re etc.; 20 pg. The abundance of each element is calculated from the gamma ray counts of several different energy peaks emitted from the same nuclides. Reference samples are made from ordinary iron meteorites, which are melted once in a high vacuum. Allende powder samples, which are crushed and powdered into fine grains of less than 20 μm in size, are also used. The reproducibility of measurements of the finest (<20 μm) fraction is so good that the fraction is useful as a reference in the determination of chemical compositions of extraterrestrial matter.

The effectiveness of these references have already been checked for homogeneity and reproducibility.

The neutron irradiations are done by a smale-scaled reactor (ex. TRIGA II) whose fluxes are $0.7 \sim 4.0 \times 10^{12}$, and also a research reactor, whose flux is 2×10^{14} (n/cm^2, sec).

After irradiation, they are removed from the capsules and mounted on polyethylene disc holders. Each sample is counted by Ge (Int.) gamma ray spectrometer for a few hours. ^{65}Ni, ^{198}Au or ^{193}Os, short-lived radionuclides, are measured by the prompt counting. More than several hundred counts are needed for determination of the nuclide with good statistics; however, sometimes it is difficult to get enough counts because of the small mass and low abundance in the spherules.

Manganese contents, determined from the photopeaks of ^{56}Mn by 1.811 keV are corrected by the contribution of the competition reaction; ^{56}Fe (n, p) ^{56}Mn. In order to estimate the amounts of this contribution, we have calculated the abundance of the Mn caused by the Fe in each standard sample using the gamma ray peak counts of ^{56}Mn and ^{59}Fe. The original Mn content of a standard sample of the iron meteorite "Canyon Diablo" is approximately 0.1 ppm in order of magnitude, and from the measurements of several samples we have obtained 5.8 ± 1.0 (μg/g) of Fe of "apparent" Mn content at the TRIGA II reactor irradiations at RSR positions. Iron contents are determined from the 1,099 and 1,292 keV peaks of ^{59}Fe. Cobalt contents are determined from the data of "famous" gammas of ^{60}Co. Nickel contents are obtained from the data at 1,482 keV of ^{65}Ni and checked by the counts at 812 keV of ^{58}Co, which is produced through a reaction of ^{58}Ni (n, p), Iridium contents are calculated from the data of 328 keV from ^{194}Ir or the counts at 315 or 468 keV from ^{192}Ir. Gold contents are obtained from the peak areas of ^{198}Au (411.8 keV); however, because of the short life (64.7 hours), the occupied counting time for each sample and the number of the samples to be irradiated at an individual irradiation study are also limited.

After several days of cooling, the activities of the long-lived nuclides, such as ^{59}Fe, ^{60}Co, ^{192}Ir etc. are measured by a low level Ge (Int.) spectrometer for a long time. The background counts in the low energy region caused by the Compton scattered gamma rays of the prominent peaks are eliminated by a low-level Ge (Int.) spectrometer with Compton suppression system.

6.4 Magnetic Stony Spherules

The chemical composition of magnetic, stony spherules gathered from rain water, ice layers and deep-sea sediments have been not so widely determined. However, high Ir, Au, Ni and Co contents indicate their extraterrestrial origin. The obtained compositions are considerably different from those of ordinary chondrites, but it can be qualitatively interpreted that cosmic meteoroids having a chondritic composition are changed into magnetic, stony spherules by thermal degeneration during their atmospheric entry. In the magnetic stony spherules, cosmic ray induced radionuclides, such as ^{53}Mn, ^{10}Be and ^{26}Al, have also been detected.

The magnetic stony spherules constitute one of the three categories of the "magnetic" spherules: [1] rough, brittle, black spherules with specific gravities lower than 4.0; [2] dull black spherules with smooth surfaces with specific gravities lower than 5.0 and [3] shiny metallic grey or shiny black spherules with occasional small vesicular cavities. These spherules have specific gravities higher than 5. The magnetic stony spherules studied here constitute the category [2]. They sometimes look just like rugby balls made of charcoal.

The magnetic stony spherules have magnetic properties, so they could be picked out from the magnetic fractions collected from ice samples and deep-sea sediments by fine tweezers under a microscope.

In 1978 Ganapathy *et al.* determined the minor and trace element concentrations of these spherules using neutron activation analysis and found that three of them had compositions similar to those of carbonaceous type 1 (C-1) chondrites. Nishiizumi (1983) deduced that the origin of the magnetic stony spherules is ordinary chondrite and they are ablated debris from chondrites by atmospheric heating, since they obtained ^{53}Mn activities similar to those of ordinary chondrites. His data could be understood on the basis of the laboratory ablation experiment that manganese atoms induced in parent chondritic bodies by cosmic rays are not ablated so highly during atmospheric heating (Notsu *et al.* 1978).

Yamakoshi (1984) determined the chemical compositions of the magnetic stony spherules from deep-sea sediments, seven individual spherules and one composite sample by instrumental neutron activation analysis of short (10 min) and long (24 hr) time irradiation in a TRIGA II reactor.

Table 6.4 shows the ^{53}Mn data obtained by neutron activation analysis. This method was already described in Section 5.4. If we use the formula:

$$N = g / \lambda\{1 - exp(-\lambda T)\},$$

where T is a cosmic ray exposure time, we can calculate the "apparent" cosmic ray exposure time, shown in the margin of Table 6.4.

The chemical compositions of magnetic stony spherules are shown in Table 6.5, which were determined by instrumental neutron activation analysis. The samples were collected from deep-sea sediments dredged at [11°00.7'S, 146°02.6'W] from a depth of 4912 m. The accuracy of the data is approximately ±5% derived from the reproducibility of the radiation measurements. Larger errors, mainly due to statistical counting errors (±σ), are indicated in parentheses (Yamakoshi 1984).

The chemical compositions determined here, however, do not coincide with those of C1 chondrites. In Fig. 6.10 the chemical compositions of the magnetic stony spherules are normalized to those of C1 chondrites, and the ratios are shown.

Table 6.4. ^{53}Mn in the magnetic stony spherules (individual and composite samples) obtained by Nishiizumi (1983).

Sample code	Weight (μg)	Number of spherules	Mn (%)	Fe (%)	Co (%)	Ni (%)	dpm ^{53}Mn /kg iron	EXPOSURE AGE (Ma)*
BKK-2	1077	1	0.31	23.5	—	0.62	202 ± 44	0.61 ± 0.13
LJSM-1	4477	72	0.21	27.2	0.05	0.62	243 ± 18	0.74 ± 0.05
LJSM-2	967	22	0.18	31.9	—	0.70	225 ± 16	0.68 ± 0.07
LJSM-3	911	23	0.55	31.0	—	0.83	262 ± 42	0.80 ± 0.16
LJSM-4	877	80	0.20	35.7	—	0.97	210 ± 37	0.63 ± 0.18

*The calculation of the exposure ages was done by Yamakoshi (1984) using the data by Nishiizumi (1983).

Table 6.5. The chemical composition of the deep sea, chondritic spherules. (Yamakoshi 1984).

Sample	Weight (μg)	Al (%)	Ca (%)	Mg (%)	Mn (%)	Ti (%)	V (ppm)	Fe (%)	Ni (%)	Co (%)
A	338	5.8	0.84 (±0.08)	14.1	0.22	0.49	187	25.9	0.53 (±0.034)	0.033 (±0.006)
B	124	3.3	2.5 (±0.3)	14.2	0.18	1.03	196	27.3	1.58 (±0.087)	0.094 (±0.002)
C	69	3.0	4.0 (±0.4)	14.8	0.34	2.20	465	30.5	0.90 (±0.08)	0.063 (±0.001)
D	215	5.0	1.7 (±0.2)	13.9	0.15	0.49	236	27.8	0.81 (±0.049)	0.020 (±0.001)
E	89	3.2	3.0 (±0.3)	14.2	0.19	0.87	262	35.5	0.77 (±0.061)	0.055 (±0.001)
F	72	3.2	3.1 (±0.3)	15.6	0.33	2.20	276	32.1	0.47 (±0.032)	0.064 (±0.001)
G	75	4.6	4.7 (±0.5)	14.9	0.29	1.20	334	30.4	0.12 (±0.038)	0.028 (±0.001)
H+I+J	194	6.2	6.6 (±0.7)	16.8	0.27	1.43	289	25.1	0.20 (±0.025)	0.024 (±0.001)

Table 6.5. (continued).

Sample	Ir (ppm)	Au (ppm)	Cr (ppm)	La (ppm)	Sm (ppm)	Eu (ppm)	Lu (ppm)	Se (ppm)
A	0.15	1.8	2320	11.1	1.3	2.52	1.3	8.2
	(±0.035)	(±0.15)		(±4.2)	(±0.29)	(±0.58)	(±1.3)	
B	0.61	5.8	3020	7.8	1.3	4.19	1.3	11.7
	(±0.079)			(±1.91)	(±0.43)	(±0.97)	(±0.57)	
C	0.12	1.5	7640	17.0	4.3	5.70	0.94	11.3
	(±0.072)	(±0.18)		(±3.2)	(±0.47)	(±0.97)	(±0.50)	
D	0.04	1.7	2720	6.0	1.5	4.10	0.19	9.7
	(±0.032)			(±1.1)	(±0.20)	(±0.50)	(±0.17)	
E	0.54	1.3	3520	10.4	2.4	5.59	0.33	8.3
	(±0.07)			(±0.90)	(±0.19)	(±1.05)	(±0.019)	
F	0.06	1.2	4830	30.0	3.9	6.79	0.42	11.7
	(±0.043)	(±0.07)		(±0.27)	(±2.3)	(±1.13)	(±0.22)	
G	0.06	2.1	3160	48.0	1.6	8.14	0.46	16.2
	(±0.053)				(±0.07)		(±0.15)	
H+I+J	0.41	0.43	1360	9.9	1.1	2.47	0.23	8.9
	(±0.033)			(±0.62)	(±0.08)	(±0.54)	(±0.087)	

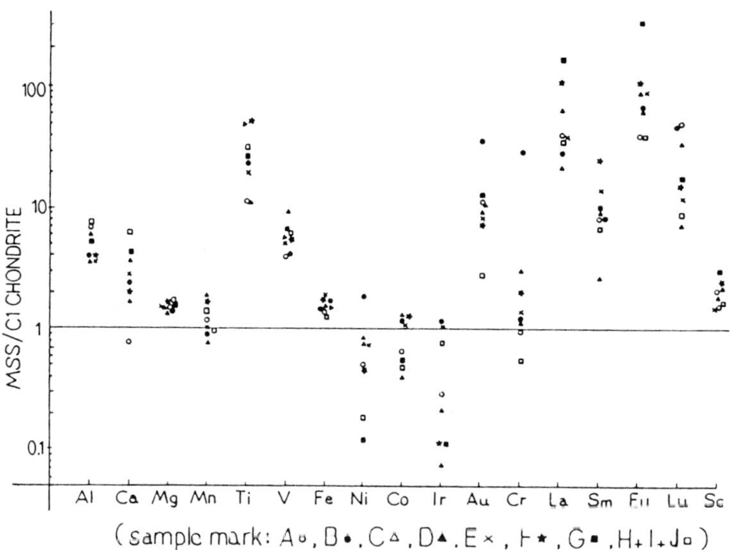

Fig. 6.10. Chemical compositions of magnetic stony spherules normalized to those of C1 chondrites (Yamakoshi 1984).

Alkali elements, such as K and Rb, are so strongly depleted in the magnetic stony spherules, that the depletion factors of these elements relative to C1 chondrites are 0.020~0.58. In Fig. 6.11, lithophile elements of individual spherules obtained by the isotope dilution method are shown, which are normalized to C1 chondrites. In the figure spherules #A and #E appear enriched in lower side REE (rare earth elements); however, there are also large negative Ce anomalies and minor negative Eu anomalies. In addition, spherule #A shows a large depletion of Ca with chondritic levels and other refractory lithophiles. Spherule #K also appears enriched in lower side REE; however, the gradual decrease from Ce to Lu superimposed a positive Eu anomaly. On the other hand, spherule #F has a nearly flat pattern (ca. 1.8 times C1 chondrite data) with no specific anomalies, even no Ce.

In some spherules, the ratios of lower part to higher part REE are relatively large, and negative Ce anomalies are observed. Such fractionations could originate from the solid/liquid partitioning of REE. If these fractionations are of pre-atmospheric origin, their parent meteoroids might be differentiated materials; however, differentiated meteorites so far reported do not show such a large negative Ce anomaly. So we can say that the depletion of Ce might have originated under oxidizing conditions when the samples entered the Earth's atmosphere.

However, the enrichment of lower part REE relative to the enrichment of higher part REE appears to conflict with the elemental volatilities. It is shown that stony spherules observed in deep-sea sediments have been affected by glass and seawater interaction. In order to make clear the alteration effects in deep sea-water, further study such as detailed SEM observation for minerals may be required. In any case, REE characteristics of the stony spherules suggest that unfractionated, very primitive materials, such as spherule #F, have existed in deep-sea sediments for a long time (Misawa *et al.* 1989).

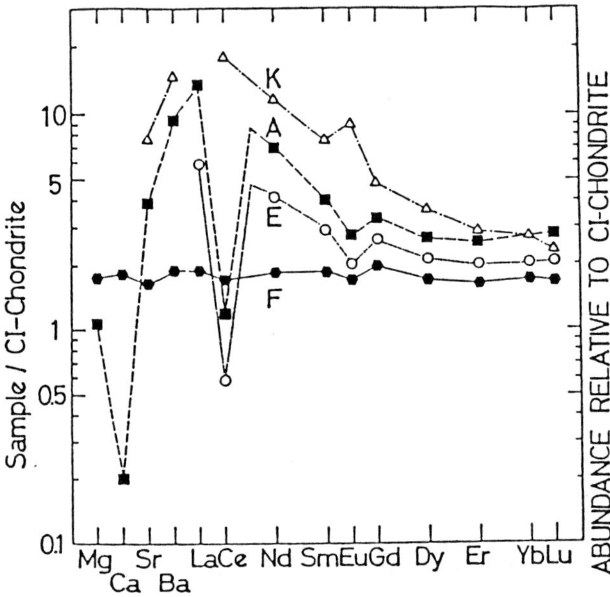

Fig. 6.11. Lithophile elemental abundances determined by the isotope dilution method (Misawa *et al.* 1989).

6.5 Stony Glassy Spherules

We have several criteria of cosmic origin for spherules, such as determinations of high contents of native iron, noble metal and siderophile components, cosmogenic radionuclide detections and also isotopic excesses of accumulated, cosmogenic stable nuclides. In addition to the above criteria, there is one more judging method: in the chemical compositions flat patterns of REE (rare earth elements), which are normalized by REE contents in C1 chondrites, can be considered.

REE are such refractory elements that they will survive frictional heating in the upper atmosphere. However, it is known that some igneous rocks also possess flat patterns of normalized REE, just like cosmic meteoroids. Such igneous rocks are considered to be brought from terrestrial lower mantle layers.

In earlier works (Nagasawa *et al.* 1979, Nagasawa *et al.* 1980-B), many glassy silicate spherules separated from deep-sea sediments were analyzed by INAA for REE and Sc, Co, Fe, Cr, Mn, Hf, Na, K, Ni, Ir and Au. Some of them, mostly colorless and transparent, showed uniform enrichment of REE and Se to about 2~7 times average chondritic values, except Ce.

The more volatile elements such as K, Na, Mn, and Cr, and the siderophile elements, such as Fe, Co, Ni and Au, are mostly depleted by an order of a magnitude or more compared with the chondritic values.

The uniform enrichment of the refractory elements and depletion of Ce and the volatile elements indicate that the spherules were originally chondritic in composition and that the depleted elements were lost away by evaporation by heating under relatively high oxygen fugacity, possibly by the frictional heating at the time of entering the terrestrial atmosphere.

The other portion of the darker coloured spherules showed higher siderophile element concentrations. These spherules showed more fractionated REE abundances relative to the chondritic values. A few samples showed a very large depletion of Ce, yet showed volatile element concentrations (Mn, Fe and Cr) much higher than those in the "chondritic" spherules mentioned above. In Fig. 6.12 REE abundance patterns normalized by C1 chondrites are shown (Nagasawa *et al.* 1980).

Thus, the Ce depletion in these spherules cannot be attributed to the process that produced Ce depletion in the "chondritic" spherules, namely the evaporation at the time of entry to the terrestrial atmosphere.

A pale-green transparent spherule and DSDP (Deep Sea Drilling Project) microtektites show much higher and more fractionated REE abundance patterns.

The REE patterns are essentially similar to those of average crustal rocks, thus indicating a terrestrial origin of these spherules.

Although DSDP microtektites have a fairly long burial time interval, measured by Glass *et al.* (1979) as 3.5 Ma in the ocean bottom, they do not show either enrichment or depletion of Ce relative to other REE.

It appears that Ce concentrations (relative to other REE) in the glassy spherules did not change significantly by the interaction with seawater during the burial time of tens of millions of years. Thus, the large depletion of Ce observed in some spherules suggests that the spherules are of extraterrestrial origin, and that the Ce anomalies were possibly produced before these samples entered the terrestrial atmosphere.

In an interim conclusion, REE and other trace element concentrations observed in small-sized spherical glassy grains obtained from deep-sea sediments have wide varieties of both terrestrial and extraterrestrial origins.

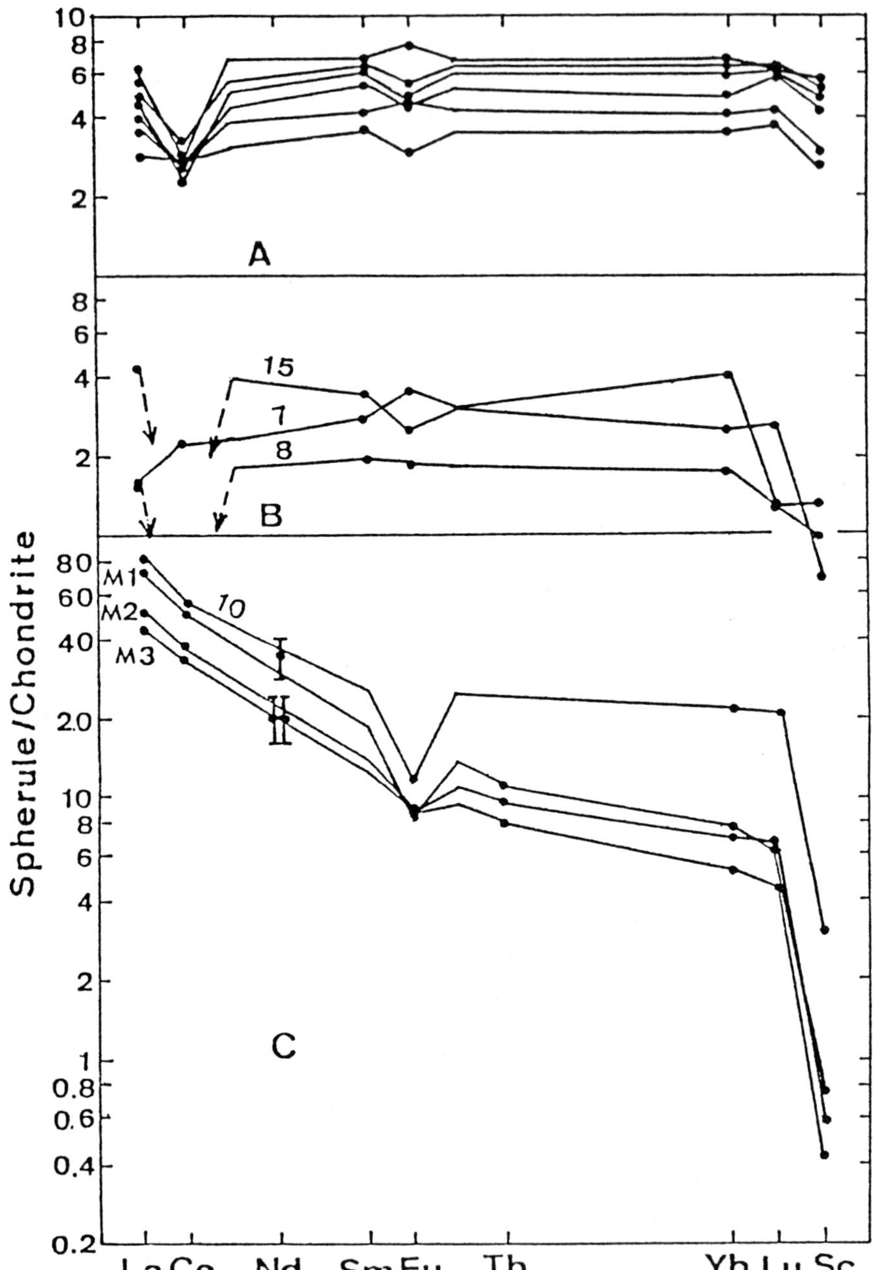

Fig. 6.12. REE and Sc abundance patterns of spherules and microtektites, (A) chondritic spherules, (B) spherules with high siderophile elemental concentrations; #7: FeO (%), 7.3: Ni (ppm), 70; Ir (ppm), 0.01. #8: FeO, 15.6; Ni, 1100; Ir, 0.25; Au, 0.02. #15: FeO, 20.2; Ni, 1500. (C) #10 and microtektites M1, M2, M3.

6.6 Spherules of Irregular Forms

From the magnetic fractions of deep-sea sediments, we can sometimes pick out spherules of irregular forms, such as gourds, dumb-bells, flasks or liquid droplets. Among the gourd-shaped particles, there are some which consist of four or five joined spheres. Therefore, there must be some "crowded" regions at higher altitudes where four or five molten iron particles can collide with each other and become such irregular forms. These forms suggest that the origins of these fused droplets spread from incoming iron meteorites. Indeed, Krinov (1964) found iron particles of similar forms in the vicinities of craters of fallen iron meteorites. Krinov found also flask and liquid droplet forms from regions where the dust-trails which had separated from incoming iron meteorites had fallen.

Therefore, these types of particles are formed from fused material which is separated continuously from the molten surface layers of falling iron meteorites in the lower atmosphere. They have no more time to be molten once again and also to be formed spherically, like rounded spherules formed in the upper atmosphere. Figure 6.13 shows some irregular grains of multi-particle adhesive structures collected from deep sea sediments (Yamakoshi 1981).

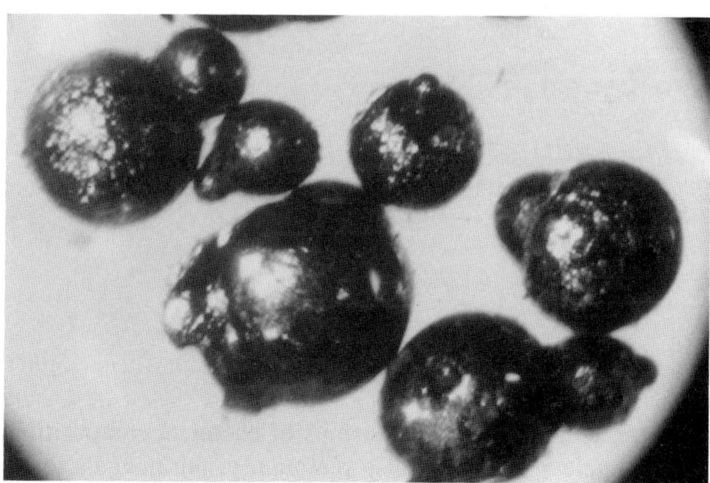

Fig. 6. 13. Irregular grains collected from deep sea sediments (Yamakoshi 1981).

Yamakoshi (1981) measured the chemical compositions of these irregularly formed particles by INAA; the results are shown in Table 6.6. In some of them, high concentrations of siderophile elements are observed; however, others are iron oxides only.

Table 6.6. The chemical compositions of irregularly formed particles taken from the magnetic fractions of deep-sea sediments.

	SAMPLE	WEIGHT (μgm)	Ni/Fe ($x10^{-2}$)	Co/Fe ($x10^{-3}$)	Ir/Fe ($x10^{-6}$)	Au/Fe ($x10^{-6}$)
GOURD SHAPED;	#15	369	0.0008	0.77	0.030	0.003
	#17	543	0.0007	3.06	0.15	0.29
	#22	123	---	0.10	0.023	0.07
LIQUID DROP SHAPED;	#13	827	---	0.002	0.002	0.0008
	#27	79	0.0014	0.0044	0.007	0.019
FLASK SHAPED; (A) typical, (B) hut shaped, (3) spherical, (a) (b) (c)	#7(A)	457	0.33	0.30	5.11	0.005
	#19(B)	310	0.002	0.043	0.02	0.028

6.7 *Analyses of Nuclear Cores and Platinum Nuggets in Cosmic Spherules*

In this section, studies on relationships of chemical composition between fusion-crust vs. nuclear cores and also platinum group nuggets in iron and/or stony spherules are described.

In the interior of iron spherules, we can sometimes find "nuclear cores" surrounded by fusion crusts made of black-magnetite materials. Brownlee *et al.* (1984) reported some nuggets whose compositions were siderophile, especially platinum group elements. PH. Bonte *et al.* (1987) studied platinum metals and microstructures in iron and in chondritic spherules. The formation processes of core-fusion crust structures and platinum nuggets in cosmic spherules may be different. Brownlee *et al.* (1984) stated the nugget-formation processes in the diagram in Fig. 6.14.

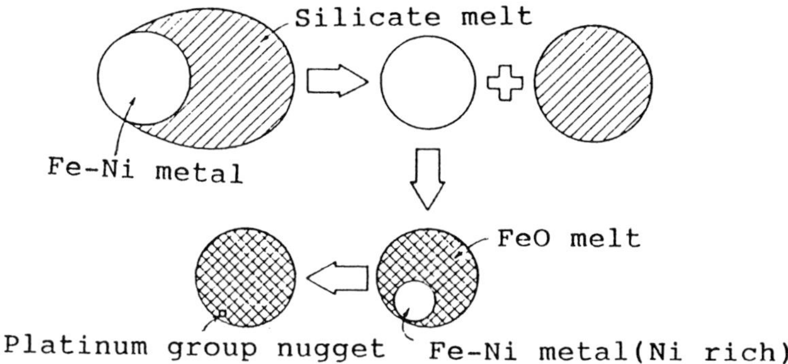

Fig. 6.14. The most "common" sequence of events that produces iron cosmic spherules (Brownlee *et al.* 1984) (They use the term cosmic "sphere", instead of "spherule").

The phenomena and phase-separation occur during atmospheric entry.

Fechtig and Utech (1964) found nuclear cores in iron and investigated nickel configuration by X-ray microanalysis. They also stated that the spherules obtained from geological layers older than the Precambrian age have no cores and are usually hollow cavities. They supposed that the nickel fraction might be replaced by other elements or nothing via chemical fractionations of deep sea-water over long period of times. They demonstrated nickel diffusion transfers into the inner part of particles by a melting laboratory experiment using a cross-section of the fusion crust of an iron meteorite, shown in Fig. 6.15.

Sasaki (1983) studied the inner core structures of a large number of iron spherules and formed an impressive picture of Fe-Ni configurations. He found also an inverse relation between the sizes of the whole spherule and the inner cores; as the whole size of the spherule becomes larger, the smaller become the cores. He moreover discovered inclusions in the core of the spherules. He stated that the core is not usually located at the center, but to one side, which was found already by Fechtig and Urech (1964); their interpretation was that during the flight of the melted iron droplets the concentrated, high density nuclear cores move slightly to the heads. Sasaki (1983) gave interesting sketches of the cross-sections of the iron spherules, which are shown below.

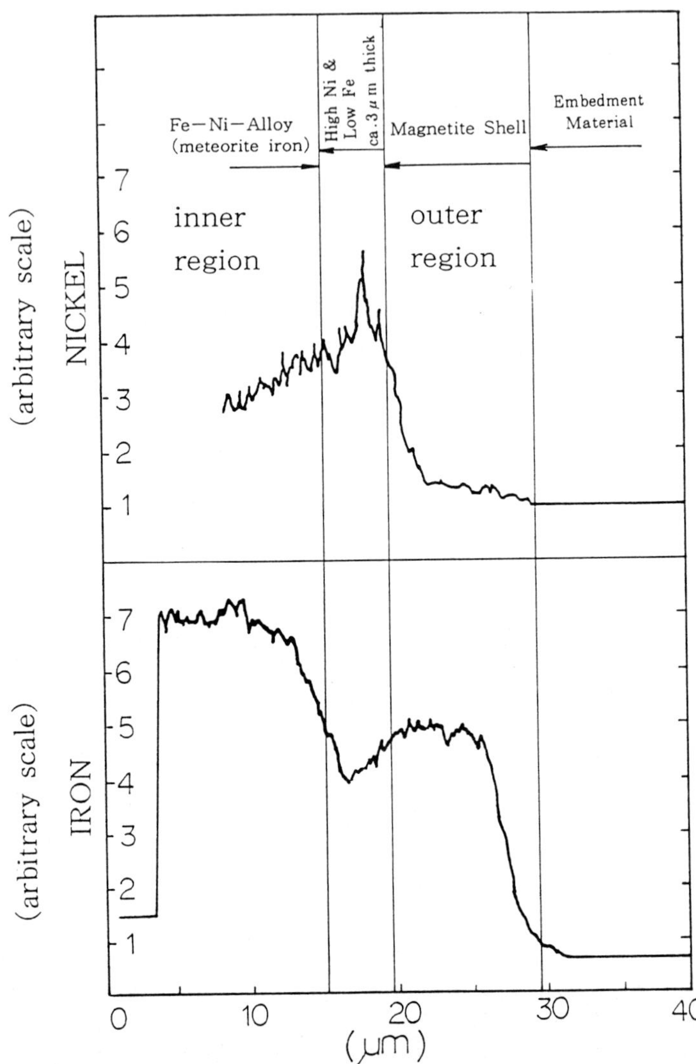

Fig. 6.15. Fe and Ni contents in the cross-section of the fusion-crust of an iron meteorite. (Fechtig and Utech 1964).

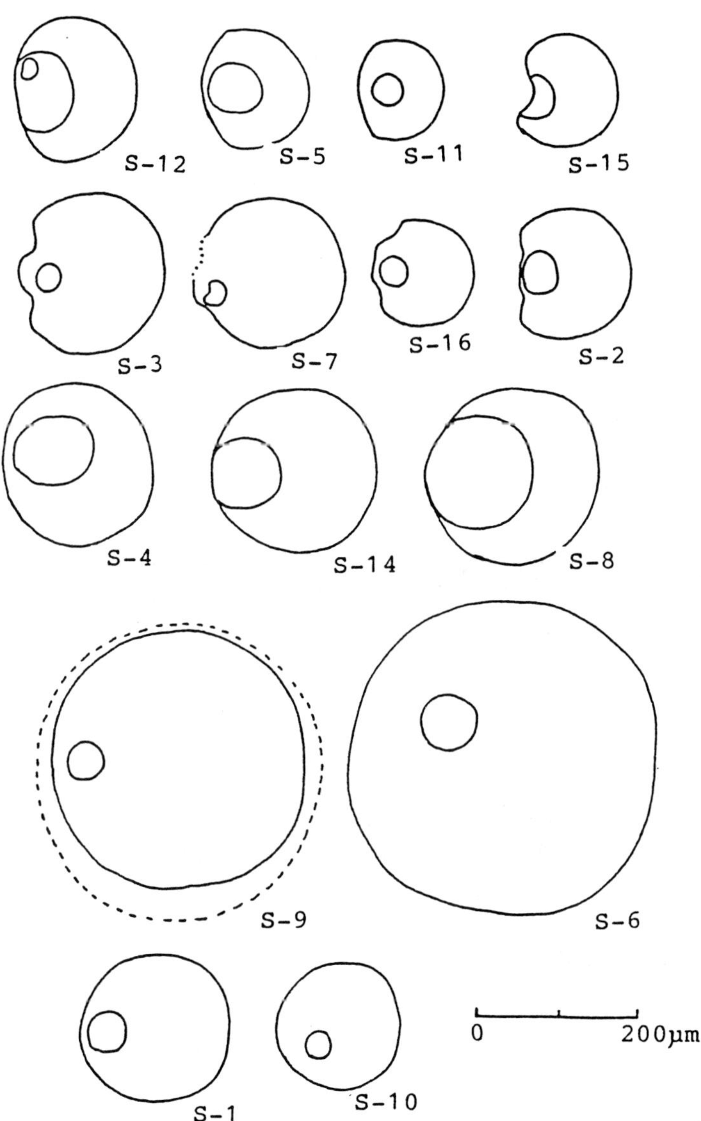

Fig. 6.16. Sketches of the polished sections of iron spherules (Sasaki 1983).

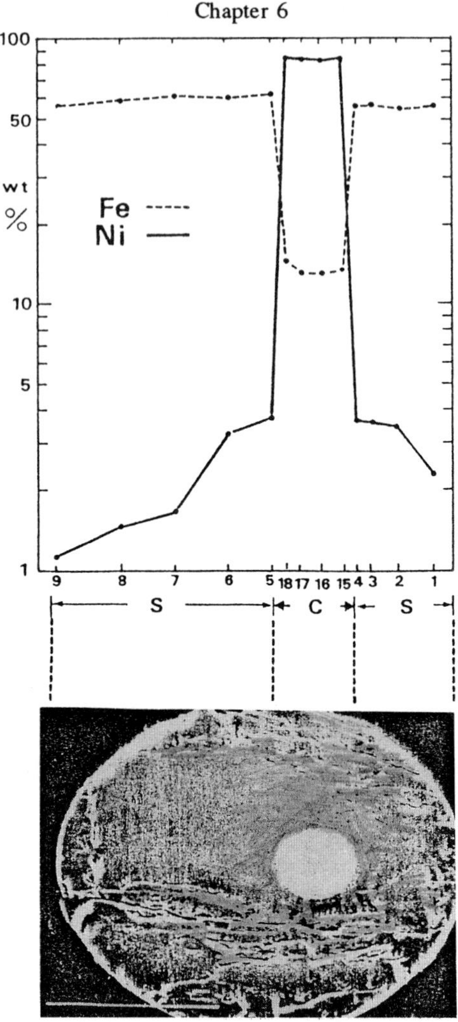

Fig. 6.17. The enrichment of nickel in and around the core, which is located fairly off center (Sasaki 1983).

We analyzed the chemical compositions of inner cores and outer crusts by INAA (Nogami *et al.* 1991), which are shown in Table 6.7.

Table 6.7. The chemical compositions of inner cores and the paired outer shells.

Sample	Weight (μg)	Fe, (%)	Ni, (%)	Co, (%)	Ir, (ppm)	Au, (ppm)	Ru, (ppm)	Os, (ppm)
core A	12.8	64.0	33.6	1.60	29.2	20.5	<40.2	<20.3
its shell	95.8	78.2	0.27	0.05	<1.3	3.6	<4.5	<2.3
[core/shell]		0.82	124	32	—	5.7	—	—
core B	40.7	57.0	44.2	2.86	27.2	2.2	<34.4	30.2
its shell	126.9	70.3	0.33	0.12	<0.17	1.0	<5.1	<2.0
[core/shell]		0.81	134	24	—	2.2	—	—
core C	47.0	53.3	46.0	1.70	42.4	7.33	—	28.3
its shell	177.0	43.3	0.16	0.044	1.3	5.5	—	—
[core/shell]		1.22	288	39	33	1.33	—	—
Canyon Diablo (reference)		92.3	7.25	0.49	2.1	2.1		3.6
Iron Spherules (5 composite ones)		75.97	4.24	0.21	5.2	0.045	—	17.0

We can see in Table 6.7 a remarkable enrichment of Ni in the core, which was first found by Fechtig and Utech (1964), and an accompanying phenomenon of concentrations of other trace noble metals in the cores of the iron spherules. The results obtained in Table 6.7 make an interesting diagram, Fig. 6. 18 (Nogami *et al.* 1991).

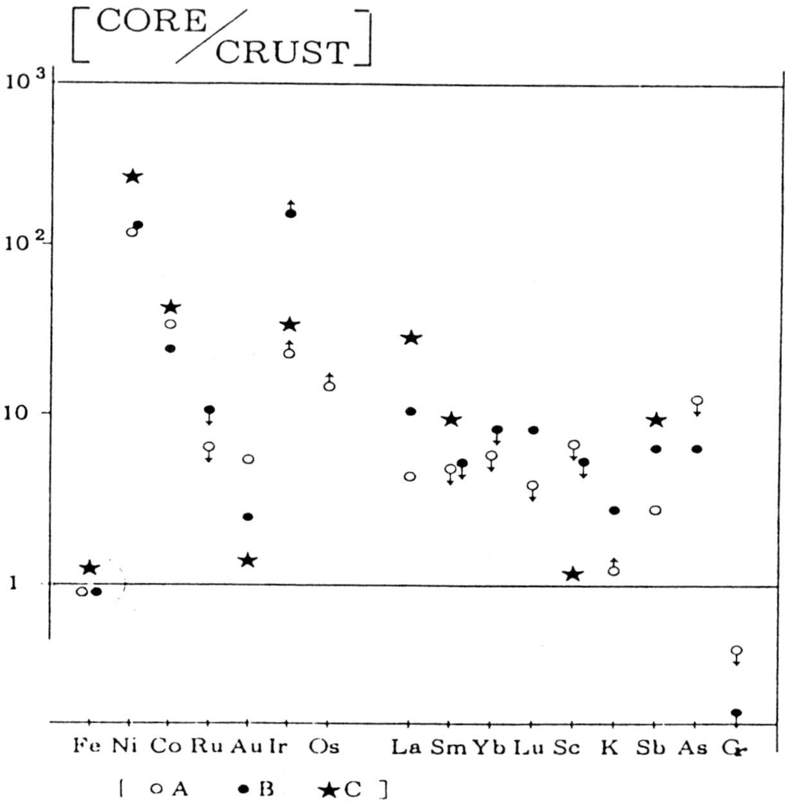

Fig. 6.18. Ratios of the elemental concentration of core and its shell (Nogami *et al.* 1991).

Brownlee *et al.* (1984) analyzed platinum group nuggets by the XMA method and showed the chemical compositions of the nuggets (Table 6.8).

Table 6.8. The chemical compositions of ten platinum group nuggets (wt%). The data are obtained by energy dispersive X ray analysis. (Brownlee *et al.* 1984).

Elements	[83H45]	[S114]	[8314]	[S-81]	[83H4]	[8319]	[8319]	[83A13]	[83A21]	[83110]
Fe	5.7	6.8	6.9	5.6	6.1	6.9	9.9	7.0	4.3	5.6
Ni	9.2	13.3	10.6	2.8	2.7	17.7	56.1	25.3	4.0	1.9
Ru	21.4	19.5	23.0	16.9	19.7	17.2	7.5	12.9	19.7	15.0
Rh	3.9	5.0	6.3	0.9	—	4.0	2.0	3.4	1.4	—
Pd	—	1.2	5.1	—	—	0.5	5.8	3.9	—	—
Os	18.6	15.0	7.4	34.2	29.9	13.6	5.5	12.0	26.0	41.4
Ir	15.6	11.9	7.2	25.9	29.4	13.5	5.4	9.9	26.0	35.2
Pt	25.8	25.7	31.6	12.0	10.0	24.7	10.3	20.6	18.5	1.8
Total	100.2	98.4	98.1	98.3	97.8	98.1	102.5	95.0	99.9	100.6

Misawa *et al.* (1989) found also a tiny nugget in a chondritic spherule using L-X ray complex spectrum; on a cathode ray image, the three elements, such as Os, Ir and Pt, are superimposed and are to difficult to separate from each other.

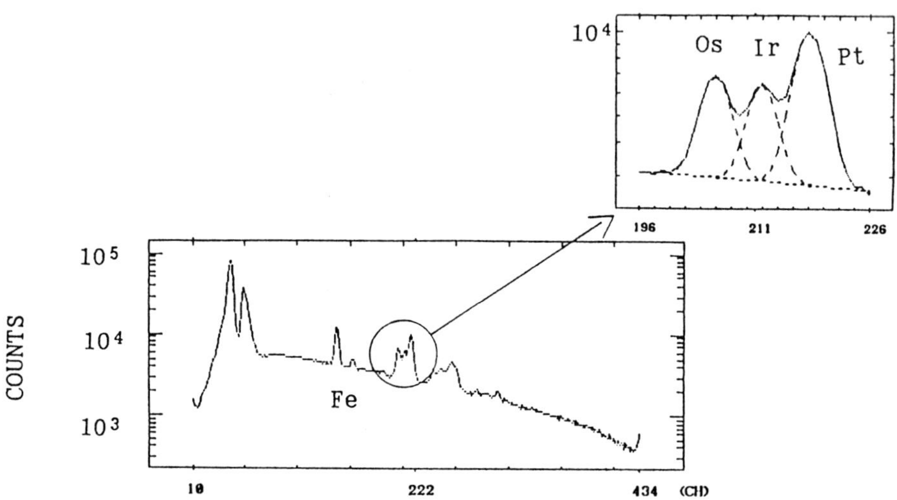

Fig. 6.19. The platinum group nugget pattern obtained in a chondritic spherule (Misawa *et al.* 1989).

The formation processes of the core-shell structure and nuggets in spherules have been not yet sufficiently studied; no success was obtained in laboratory simulation experiments. Spherule and inner structure formation processes have been not studied so much, therefore in future this problem will be very important in laboratory experiments, because we must reveal meteor-flashing phenomena in the upper atmosphere and get information about preatmospheric, chemical and isotopic compositions. These problems will be studied in Chapter 7.

6.8 Refractory Elements in Spherules

In order to study the inhomogeneity of the solar materials in their initial stage, the enrichment factors and isotopic anomalies of refractory elements in deep sea spherules relative to the average solar abundance have been examined using instrumental neutron activation analysis (INAA). Fairly high enrichment factors and isotopic anomalies of refractory elements were observed in several spherules; however, it can be said that after the ablation process of the spherules during atmospheric heatings, they were not inconsistent with normal values in the solar system.

It has been verified that during the development of the solar system, the peak temperatures were inhomogeneous throughout the solar system and that the dynamical mixing of the condensed materials was also imperfect. In particular, the chemical and isotopic compositions of elements which have considerably high condensation temperatures, the most primitive fractions found in carbonaceous chondrites and also in interplanetary dust, are thought to be good indicators for studies on the pre-history of the solar system.

In Fig. 6. 20, we compare the contents of refractory, siderophile elements in deep-sea spherules and metal grains from carbonaceous chondrites (Yamakoshi and Honma 1985). The data of stony spherules determined by Ganapathy *et al.* (1978) are also plotted.

According to the condensation theory in equilibrium, the condensed materials in the solar nebula have the same chemical compositions as those of the solar abundance.

Grossman and Ganapathy (1976) have given the mean enrichment factors as 18.6 for refractory siderophile and rare earth elements in metal grains from the Allende meteorite. However, Palme and Wlotzka (1976) and Palme *et al.* (1982) found extremely high enrichment factors of refractory elements in metal grains from carbonaceous chondrites. These very high enrichment factors could hardly be explained by the condensation theory in equilibrium, so they proposed a model of alloy condensation of the refractory elements in the early stage of the solar nebula (Palme and Wlotzka 1976).

However, we have not yet found such high factors in deep-sea spherules. These factors (~18) coincided by chance with those given by Grossman and Ganapathy (1976).

However, these spherule factors can be explained sufficiently by melting and solidification processes of micro-meteorites during the atmospheric heatings.

In Fig. 6. 20 one trial was performed. The correlation between (Ru/Os) and (Ir/Os) was examined to obtain effective clues to understand some kinds of groupings of the data. The data of these metal grains in carbonaceous chondrites as determined by the Chicago group (Kawabe *et al.* 1982) fall on trajectory (a), and the end member is the bulk composition of type 1 carbonaceous chondrites.

However, the deep-sea spherules fall on the other curve (b), whose end member is ordinary iron meteorites, e.g. Canyon Diablo. In addition, a strange family of data from the Ornans carbonaceous chondrite gathers on line (c).

This trial is indeed suggestive; however, at the present time no decisive conclusion can be drawn.

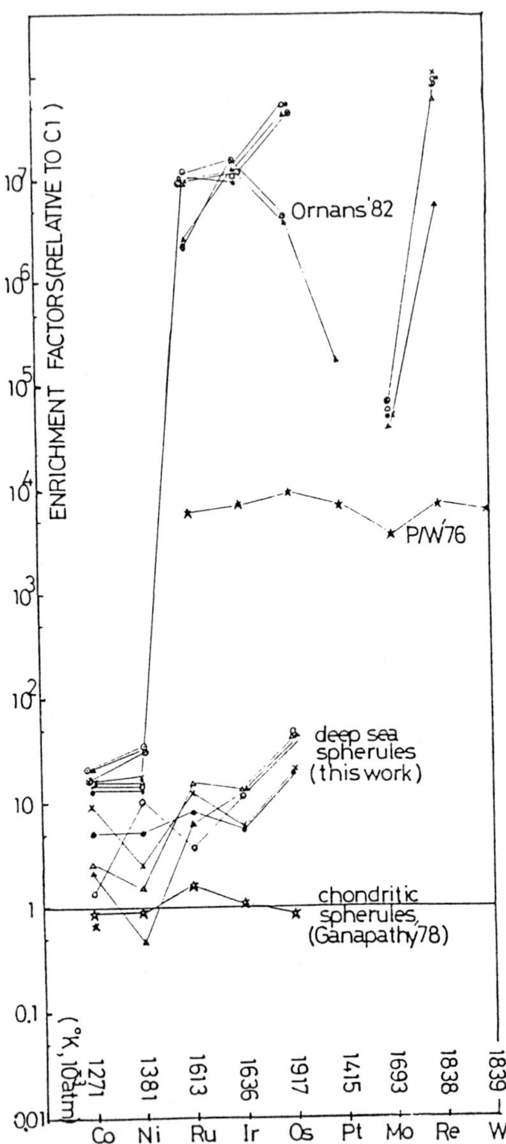

Fig. 6.20. The enrichment factors are defined as the ratio of the elemental compositions to those of type 1 carbonaceous chondrites (Palme and Wlotzka 1976, Ganapathy *et al.* 1978, Yamakoshi and Honma 1985).

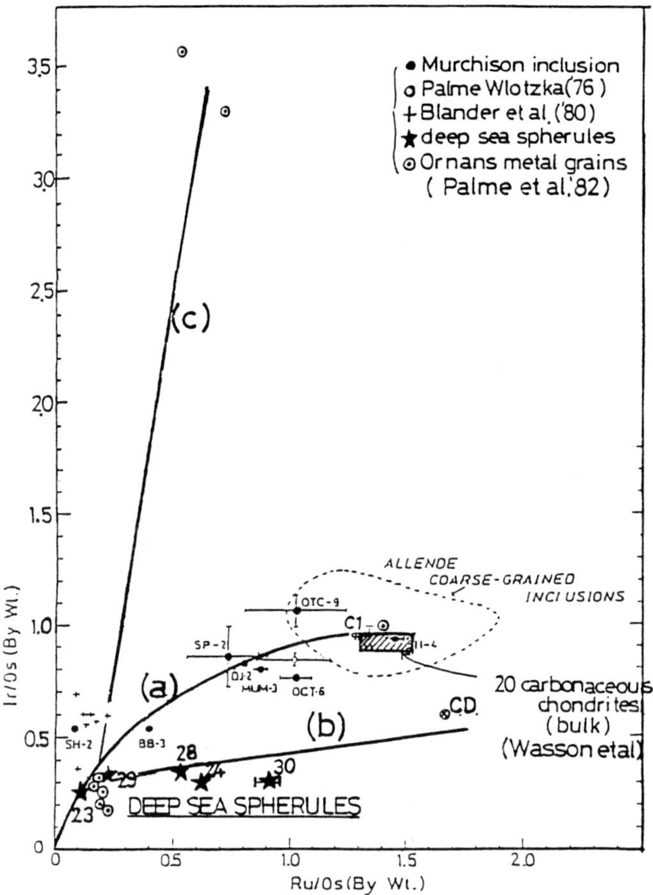

Fig. 6.21. (Ru/Os) vs. (Ir/Os) plots of the deep-sea spherules and also metal grains in many carbonaceous chondrites (Yamakoshi and Honma 1985).

Osmium and iridium are very suitable elements for studies on pre-solar and extra-solar materials. In this work, Os-184, Os-190, Os-192 and also Ir-191 and Ir-193 were measured by INAA.

Os-184 is produced through p-process only, and Os-192 is induced through r-process only in super-novae explosions. The other nuclides are produced through a mixture of both r- and s-processes.

According to Seeger *et al.* (1965), the ratio of the nuclides produced through the r- and s-processes is 7 for Os-190, 14 for Ir-191 and 22.8 for Ir-194. However, the half-lives of Os-193 (induced from Os-192) and Ir-194 (induced from Ir-193) are so short that precise measurements of these nuclides have not yet succeeded. In Fig. 6. 22, the ratios of Os-185 and Os-191 activities in deep sea metallic spherules and also in Canyon Diablo are shown.

We also measured Os reagent samples many times; the reference values are shown also. Almost all spherules are not inconsistent with the reference values; if we take $\pm 2\sigma$ as the statistical errors, these data suggest the results are not so strange (Yamakoshi and Honma 1985).

Many experimental data obtained in metal grains from carbonaceous chondrites suggest the inhomogeneity of the condensed materials in the solar nebula. However, from the deep-sea spherules we could not obtain such decisive evidence in chemical and isotopic compositions derived from the inhomogeneity of the early stage of the solar nebula.

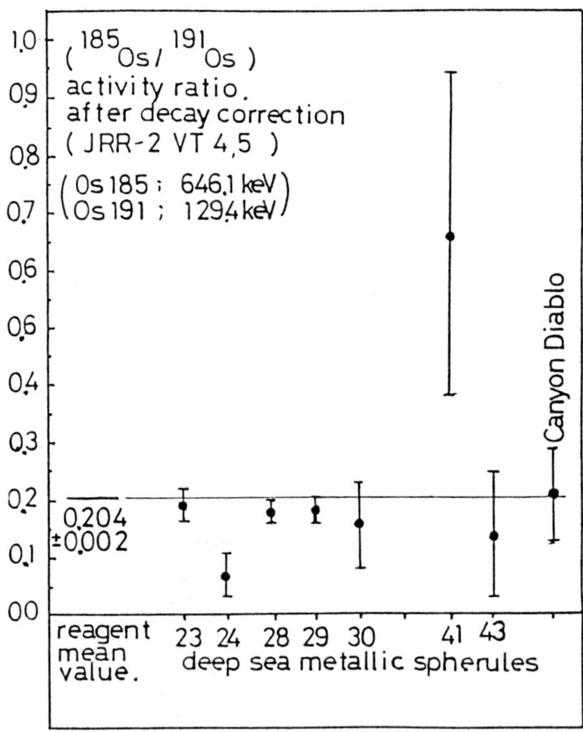

Fig. 6.22. The activity ratio of (Os-185/Os-191) in deep-sea spherules and Canyon Diablo. The reference value of the reagent is also shown (Yamakoshi and Honma 1985).

6.9 Mysterious Sample from Deep-Sea Sediments

From magnetic fractions from deep-sea sediments collected at a depth of more than 5500 m in the Central Pacific Ocean, a mysterious sample was found. It is shown in Fig. 6. 23.

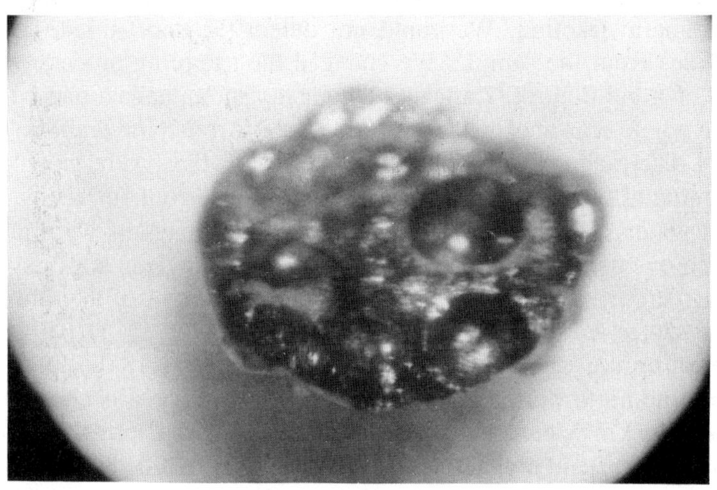

Fig. 6.23. A mysterious sample from deep-sea sediments, a spherule container of cosmic origin?

The size of this sample is ca. 2.0 × 1.5 mm, the weight is 0.562 g. INAA analysis shows the chemical composition of [Fe] = 74% and traces of Ni.

In the photograph, several cavities or microcraters of a few hundred μm in diameter can be seen, and it looks just like one side of a "spherule-container", which is made of iron oxides.

One plausible interpretation for the formation process of this sample is as follows. Iron meteoroids rushed into the Earth's atmosphere, and after the meteor-flashing, iron spherules which were separated from the molten parent bodies and already solidified hit the surface of this iron broken piece, which was also separated from the parent bodies; however, it was not yet as well solidified because of its larger size. So, several cavities might have remained on the surface of this sample.

6.10 Search for Extraterrestrial Materials in Fresh Snow

Niizeki *et al.* (1979) and Noma *et al.* (1984) tried to examine extraterrestrial materials in magnetic fractions gathered from a large volume of fresh snow, which was melted in a stainless steel bath with propane burners in 1979 and then filtered with Millipore filters of 8 μm pore size. The filtered water volume was about 28,000 liters. We obtained 28 g of ashed fraction and 1.34 g of the magnetic fraction. We could not detect the short-lived, cosmogenic radionuclides from the sample. We changed the sampling process in 1982. A roof area of a building of Yamagata University in Yamagata had been closed during the whole season of winter in 1982~1983. Around the drain-holes a large number of disc-magnets covered with polyethylene bags were mounted. When the accumulated snow layers had vanished away, the roof surface was washed with a jet water stream of a fire-hose and then the magnets were picked out. Careful removements of the magnets from the polyethylene bags gave us a lot of the magnetic fraction. We could process about 210 tons of the rain and snow during the winter season.

The sampling site is located at (38.52°N, 141.18°E) in Yamagata Prefecture, the north-northwestern part of Japan. This area is characterized by heavy snow fall in winter and was considered to be fairly free from air-borne terrestrial and artificial contaminants. We picked up rounded particles from the magnetic fraction. The total weight of the collected grains was 1.85 g, which were counted with an extremely low background counter system at Mt. Nokogiriyama.

We aimed to detect ^{56}Co (77.3 days) and other short-lived nuclides, such as ^{54}Mn (312.5 days), ^{55}Fe (2.60 y), ^{57}Co (270 days), ^{60}Co (5.26 y), which are considered to be generated from iron bombardments of solar cosmic rays in space.

In this case a cosmogenic atom number (N) in space is expressed:

$$dN \,/\, dt = g - \lambda N, \tag{1}$$

where g is the production rate of the nuclide in meteoroids per unit time, $\lambda = a$ mean life of the radionuclide. Integration of equation (1) gives

$$N = (g \,/\, \lambda)[1 - \exp(-\lambda T)], \tag{2}$$

where T is the cosmic ray exposure age. We select such a short lived nuclide that we can have $\exp(-\lambda T) << 1$. Thus, the atom number of the nuclide in meteoroids will be $N = (g/\lambda)$ at the top position of Earth's atmosphere. Because of protection from cosmic rays by the atmosphere the radionuclide decays monotonously as

$$N = (g / \lambda)\exp(-\lambda t). \tag{3}$$

When we can determine two nuclides derived from the same target (ex iron), so we have a ratio (R) of the radioactivities as follows;

$$R = \frac{\lambda_1 N_1 D}{\lambda_2 N_2 D} = \frac{g_1 \exp(-\lambda_1 t)}{g_2 \exp(-\lambda_2 t)} \tag{4}$$

where D is the weight of the sample and t is a common accretion time of meteoroids involved in the magnetic sample, t is a residence time in the atmosphere, or a dwelling time of meteoroids in the atmosphere.

In Fig. 6.24 a schematic diagram of the time sequence of the falling time, the sampling time and the counting time in this study is shown.

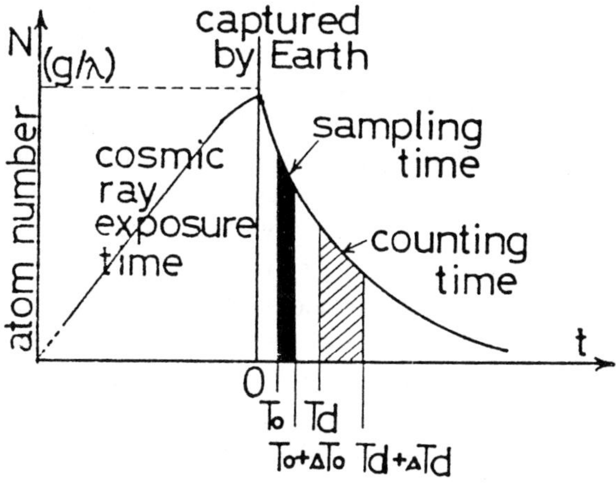

Fig. 6.24. The schematic diagram of the time sequence.

From equation (4) we have the dwelling time, $t = T_0$:

$$T_0 = [1 / (\lambda_1 - \lambda_2)] \cdot \ln[(g_2 / g_1) \cdot R], \tag{5}$$

A detailed expression of equation (5) is given as follows:

$$T_0 = \left[1/(\lambda_1 - \lambda_2)\right] \cdot \ln\left[(g_1 / g_2) \cdot (\lambda_2 / \lambda_1) \cdot (\eta_1 / \eta_2) \cdot (C_2 / C_1) \cdot \{1 - \exp(-\lambda_1 \Delta Td)\} / \{1 - \exp(-\lambda_2 \Delta Td)\}\right], \qquad (6)$$

where ΔTd = the counting time interval and η = the counting efficiency due to the gamma ray energy.

Now we choose ^{56}Co as nuclide (1) and ^{54}Mn as nuclide (2). After a simple calculation $g_1 = 2.93$ and $g_2 = 0.39$ (dpm/g dust). $\Delta Td = 32.9$ days. $\eta_1 = \eta_2 = 3.4\%$ (both gamma rays are nearly equal!), $C_1 = 119 \pm 62$, $C_2 = 261 \pm 75$. So we have $Td = 397$ days, therefore $T_0 = 260$ days ($\Delta T_0 = 98$ days) as the results.

The residence time of fine grains in the atmosphere could be estimated roughly to be 260~360 days using the cosmogenic nuclide measurement method. However, the value is not so inconsistent compared with the falling time of fine grains studies in meteorology.

6.11 Cosmic-Ray Exposure Age Determinations of the Spherules and Discussions of Their Origins

In this section the cosmic ray exposure ages of the cosmic, metallic spherules gathered from deep-sea sediments are determined with various experimental methods, such as low-level countings (Ni-59), neutron activation analysis (Mn-53), high-energy accelerator mass spectrometry (Be-10 & Al-26) and mass spectrometry (K-40). Exposure ages of 0.3~50 Ma were obtained.

According to the Poynting-Robertson Effect, the starting points of the micro-meteorites (parent bodies of the spherules) are located at inner regions of the orbits of Saturn.

The cosmic ray exposure age [T] is calculated by the following two methods. [1] The amount of accumulated cosmogenic stable nuclides (including extremely long-lived radionuclides, such as ^{40}K) (S) is divided by the production rate (R), so we have

$$T = S / R.$$

[2] The activity ratio (A) of two long-lived radionuclides induced from the same target materials, such as ^{56}Fe, by cosmic rays is obtained in the following formula:

$$A = \lambda_1 N_1 / \lambda_2 N_2 = (g_1 / g_2) \cdot \{1 - \exp(\lambda_1 T)\} / \{1 - \exp(1 - \exp(\lambda_2 T)\},$$

where g = the production rate of the nuclide at 1 AU from the sun and λ = the mean life of the nuclides. And the "common" age (T), so to say the averaged exposure age, can be calculated. Here, we have $g = n\sigma f$, where n = the target atom number, σ = the production cross-sections and f = the cosmic ray intensity.

In the case of solar cosmic rays, the flux is dependent on the radial distance from the Sun. However, we can cancel out the radial factors from the ratio of (g_1/g_2). Therefore, the cosmic ray exposure age (T) is free from the origin, motion, source positions and trajectories of the dust in interplanetary space.

In the case of high energy Galactic cosmic rays, the flux is approximately constant except in the region closest to the Sun. Therefore, the ratio of (g_1/g_2) is also independent of the dust motion and orbits in space. The obtained results are summarized in Table 6.9.

In the first experiments, we measured ^{59}Ni and ^{53}Mn in the nickel and manganese fractions extracted chemically from a large volume of deep-sea sediments. Therefore, we did not see any meteoritic materials contained in these sediments.

We obtained a "mean" value of the cosmic ray exposure age to be 0.32 Ma. In this experiment we did not discuss the size, composition, colour, density of the spherules, so to say, we have skipped to recognizing "cosmic matter" as a "black box" (Yanagita *et al.* 1979).

In the second study, the amounts of the cosmogenic radionuclides were determined by AMS (Accelerator Mass Spectrometry) of the University of Pennsylvania (Raisbeck *et al.* 1982).

In the third experiment, the cosmic ray exposure age of the metallic spherules in several size groups were studied in order to obtain information about the dust motion in the interplanetary space due to the Poynting-Robertson effect. However, manganese nuclides induced in iron parent bodies by nuclear reactions are liable to evaporate away from the parent bodies due to atmospheric heating (Nogami and Yamakoshi 1985). Therefore, we could get only apparent ages (Yamakoshi *et al.* 1982).

In the fourth study, the cosmic ray exposure ages were estimated through excess amounts of ^{40}K induced by spallogenic reactions due to Galactic cosmic ray using a highly sensitive mass spectrometer (Kobayashi *et al.* 1982).

According to the Poynting-Robertson Effect, the cosmic ray exposure age $[T]$ is given by the formula:

$$T = 700 \cdot ad(r^2 - 1)\,yr,$$

where a = the size of the dust, d = the density of the dust and r = the starting position of the parent bodies in space (AU) from which the dust is removed and for the first time irradiated by the cosmic radiation in the solar system. Of course, the size of the spherules is not the same as the pre-atmospheric, original size of the dust. Therefore, we take the spherule size as the lower limit, so that the cosmic ray exposure time will be defined as the upper limit. The data used in this work suggest that the all spherules which have been examined already have originated from the inner region of the orbit of Saturn (Yamakoshi 1990).

Table 6.9. The obtained results of the cosmic ray exposure ages and the starting positions of the dust, from which it is exposed to cosmic rays at the first time is estimated due to the Poynting-Robertson Effect. (Yamakoshi 1991).

Code[Spherule Size] (μm)	[Nuclide Pair]	[Exposure Age] (My)	[Starting Position] (AU)	[Authors]
(1) total pelagic sediments	^{59}Ni & ^{53}Mn	0.32	2.6 (mean S = 2.4 μm)	Yanagita, *et al.* (1979)
(2) metallic 180~330	^{10}Be & ^{26}Al	3.9	2.0 (mean S = 240 μm)	Raisbeck *et al.* (1982)
(3) metallic 180~260	^{59}Ni & ^{53}Mn	~0.01 *	~1.0	Yamakoshi *et al.* (1982)
(4) metallic				
360	K-isotopic excess	21 ± 15	3.4	Kobayashi *et al.* (1982)
340		3.4 ± 2.9	1.7	
300		19 ± 8	3.5	
300		52 ± 23	5.7	
(5) 341	^{10}Be & ^{26}Al	0.01	~1.0	Nishiizumi (1989)

*Due to the ablation loss of Mn nuclides from iron phase of meteoroids during atmospheric heating.

REFERENCES

Blanchard M. B., Brownlee D. E., Bunch T. E. Hodge P. W. and Kyte F. T., 1980 Earth Planet. Sci. Lett. 46 178.

Brownlow A. E., Hunter W. and Parkin D. W., 1966 Geophys. J. Roy. Soc. 12 1.

Brownlee D. E., Bates B. A. and Wheelock W. W., 1984 Nature 309 693.

Cassidy W. A., 1964, Ann. New York Acad. Sci. 119 318.

Dohnanyi J. S., 1978 Particle Dynamics, Cosmic Dust ed. J. A. M. MacDonnell (John Wiley & Sons; New York) Chap. 8 527.

Fechtig H. and Utech K. 1964, Ann. New York Acad. Sci. 119 243.

Fireman E. L. and Kistner G. A., 1961, Geochim. Cosmochim. Acta 24 10.

Ganapathy R., Brownlee D. E. and Hodge P. W., 1978, Science 201 1119.

Gruen E. T., Zook H. A., Fechtig H. and Giese R. H., 1985, ICARUS 62 244.

Kobayashi K., Ohki M. and Ohashi H., 1982, Abstracts Internrn Conf. Geochronology (Nikko) 186.

Krinov E. L., 1964, Ann. New York Acad. Sci. 119 224.

Laevastu T. and Mellis O., 1955, American Geophys, Union 36 385.

Laevastu T. and Mellis O., 1961, J. Geophys. Res. 66 2507.

Larson R. R., Dwornik J. and Adler I., 1964, Ann. New York Acad. Sci. 119 282.

Marvin U. B. and Einaudi M. T., 1967, Geochim. Cosmochim. Acta 31 1871.

Maurette M., Jehanno C., Robin E. and Hammer C., 1987 Nature 328 699.

Merrihue C., 1964, Ann. New York Acad. Sci. 119 351.

Millard H. T. Jr. and Finkelman R. B., 1970, J. Geophys. Res. 75 2125.

Misawa K., Ma Shulan, Yamakoshi K., Nogami K. and Nakamura N.,1989, Abstracts Lunar Planet. Sci. Conf. XX (Houston) 695.

Misawa K., Yamakoshi K., Nogami K., Yamamoto K. and Nakamura N., 1989, Abstracts of 14th Antarctic Meteorites Symposium 59-1.

Murray J. and Renard A. F. 1891, Report Challenger Exped. Deep Sea Deposits Chap. IV.

Murrell M. T., Davis P. A. Jr. and Nishiizumi K., 1980 Geochim. Cosmochim. Acta 44 2067.

Nagasawa H., Yamakoshi K. and Shimamura T. 1979, Geochim. Cosmochim. Acta 43 267.

Nagasawa H.., Yamakoshi K. and Higuchi H., 1980, Geochem. J. 14 1.

Nishiizumi K., 1983 Earth Planet. Sci. Lett. 63 223.

Nogami K., Shimamura T., Tazawa Y. and Yamakoshi K., 1980 Geochem. J. 14 11.

Nogami K. and Yamakoshi K., 1985, Geochem. J. 19 97.

Nogami K., Omori R. and Yamakoshi K.,1988 Bull. General Educ. Dokkyo Univ. Sch. Medic. 11 59.

Nogami K., Misawa K., Omori R., Jianguo Ma and Yamakoshi K., 1991, Proc. #128 IAU Colloquium "Origin and Evolution of Interplanetary Dust" ed. A. C. Levasseur-Reguord and H. Hasegawa Kluwer Acad. Publish. Tokyo pp 109.

Nordenskioeld A. E., 1874 described by J. Murray in Report Challenger Exped. IV.

Notsu K., Onuma N., Nishida N. and Nagasawa H., 1978, Geochim. Cosmochim. Acta 42 903.

Oepik E. J., 1955, Nature 176 926.

Parkin D. W. and Tilles D., 1968, Science 159 936.

Pettersson H. and Rotschi H., 1951, Nature 166 308.

Ph Bonte *et al.* 1987 J. Geophys. Res. 92 B4 E641.

Raisbeck G. M., Yiou F., Klein J., Middleton R., Yamakoshi K. and Brownlee D. E. 1983 Abstracts 14th Lunar Planet. Sci. Conf. (Houston) 621.

Sasaki T.,1983 Education of Geosciences in Osaka 5 9 (in Japanese).

Schmidt R. A. and Keil K., 1966 Geochim. Cosmochim. Acta 30 471.

Smales A. A., Mapper D. and Wood A. M. 1957, Analyst 82 75.

Shimamura T., Arai O. and Kobayashi K. 1977, Earth Planet. Sci. Lett. 36 317.

Shimamura T., Yanagita S., Yamakoshi K., Nogami K., Arai O., Tazawa Y. and Kobayashi K., 1979, Earth Planet. Sci. Lett. 42 379.

Yamakoshi K., Nogami K. and Shimamura T., 1981, J. Geophys. Res. 86 #B4 3129.

Yamakoshi K., Ohashi H. and Imamura M., 1982 Abstracts of Intern. Conf. Geochronology (Nikko) 401.

Yamakoshi K., 1984, Geochem. J. 18 147.

Yamakoshi K., 1985 Properties and Interactions of Interplanet. Dust #85 IAU Colloquium ed. R. H. Giese and P. Lamy (D. Reidel).

Yamakoshi K., 1985, J. Geomagnet. Geoelectr. 37 205.

Yamakoshi K., 1991, J. Phys. G. Supplement of 14th Europhys. Conf. (Bratislava)

Yamakoshi K. and Nogami K., 1991, Proc. #126 IAU Colloquium (Kyoto) ed. A. C. Levasseur-Regourd and H. Hasegawa (Kluwer Academic Publish. Tokyo) pp. 105.

Yanagita S., Yamakoshi K, and Imamura M., 1979 Proc. Intern. Conf. Cosmic Rays (Kyoto) OG-12-20.

Chapter 7

Simulation Test Experiments for Dust Characteristics Studies

7.1 Studies on Changes of Elemental Abundance in Cosmic Matter by Ablation Experiments

When cosmic stony and/or iron materials are melted by heating in the upper atmosphere, it is very important to examine the features of the individual elements in the parent bodies as well as in the meteor-flashing process.

Behaviors of elements in stony materials

Notsu *et al.* (1978) examined the volatilization features of the Allende meteorite, in which Ca and Al rich inclusions (CAI) were considered to originate from early condensates of the solar nebula at high temperature or from evaporation residues of precursor-dust in the early solar system. They heated Allende bulk samples in a vacuum and analyzed the residual fraction (RF) at various stages. Their interesting results showed that the chemical compositions of a chondrule in Allende are similar to those of the residue in 65% of RF. Furthermore, the chemical composition of a CAI is also similar to that of the residue in 4% of RF.

They performed the evaporation experiments in two ways, by heating the bulk Allende samples by DC arc and by resistance heating in a vacuum.

They attained temperatures of 2000°C by the DC arc method and 1300~2000°C by the resistance heating method. The chemical compositions of various stages of the residual fractions (RF) were determined by instrumental neutron activation analyses (INAA) and by an electron probe microanalyzer (EPMA).

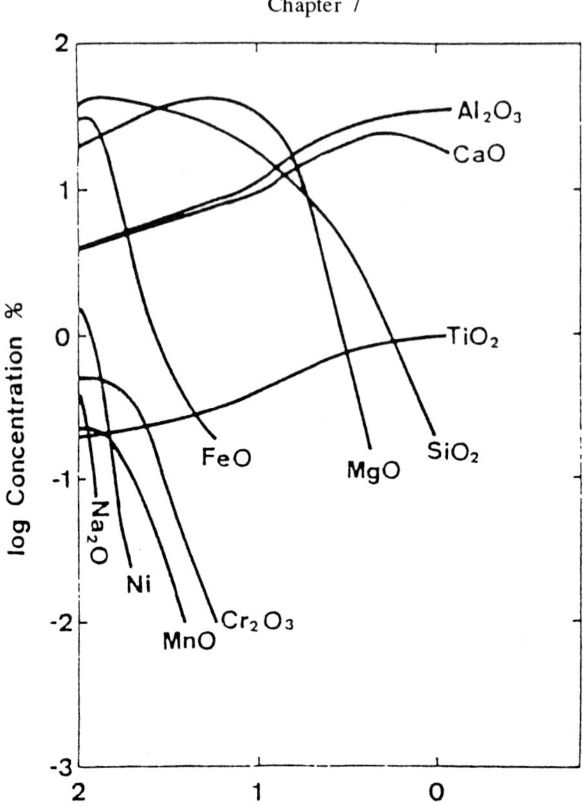

Fig. 7.1. The obtained major elemental features determined by the DC arc method are shown (Notsu *et al.* 1978).

They interpreted the thermal behaviors of the elements as a sequence of the evaporation;
[Na]→[Fe, Ni, Co, Au, Ir]→[Fe, Mn, Cr]→[Mg, Si]→[Al, Ca, Ti, Sc, REE], (V; abnormal).

Their experiments were carried out in a not so high vacuum, so the behaviors of the elements were influenced not only by the thermal degeneration, but also by oxidization of the elements. If oxidization did occur, the sample weights should be changed in appearance, and the "true" residual fractions would never be determined. In their experiments, siderophile elements were rapidly evaporated away, but Ca, Al and REE tended to accumulate in the samples.

An application check involved the chemical compositions of magnetic, stony spherules (MSS), which are produced by ablation during passage in the Earth's atmosphere (Yamakoshi 1984).

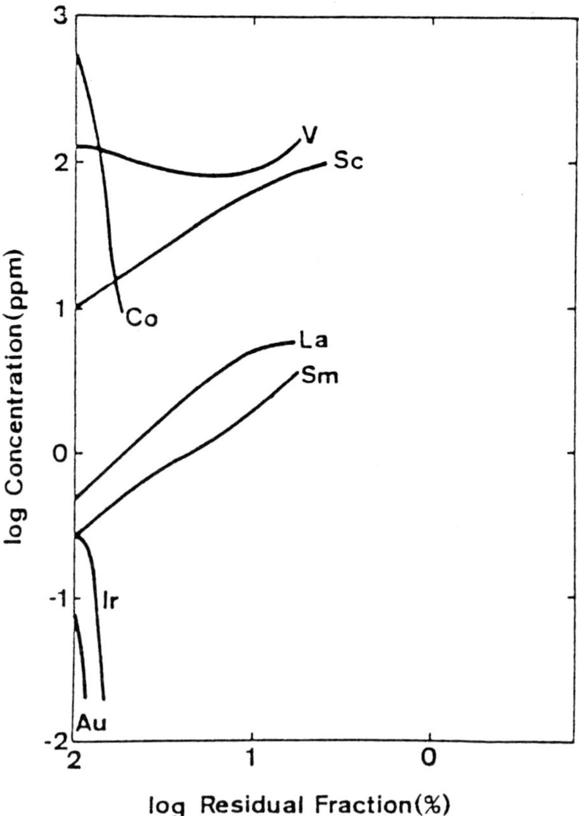

Fig. 7.2. The obtained minor and trace elemental features determined by the DC arc method are shown (Notsu *et al.* 1978).

Among major elements, Al, Ca, Mg, Ti, Fe and REE are enriched, while Ni, Co and Ir are depleted. All data except Au* can be interpreted qualitatively (more or less) by the results obtained in laboratory experiments.

The comparison of the results obtained in simulation with the MSS data normalized by the solar abundance could bring us much detailed information about the thermal process, such as vacuum pressure or altitude intervals of meteor flashing, temperature, duration time interval etc. More detailed laboratory experiments with multi-parameters must be done in the near future.

*The behavior of Au in meteoroids is peculiar. The concentration of Au is abnormal and overabundant in many cases, when samples are collected in the terrestrial atmosphere and deep-sea sediments (see Chapter 6).

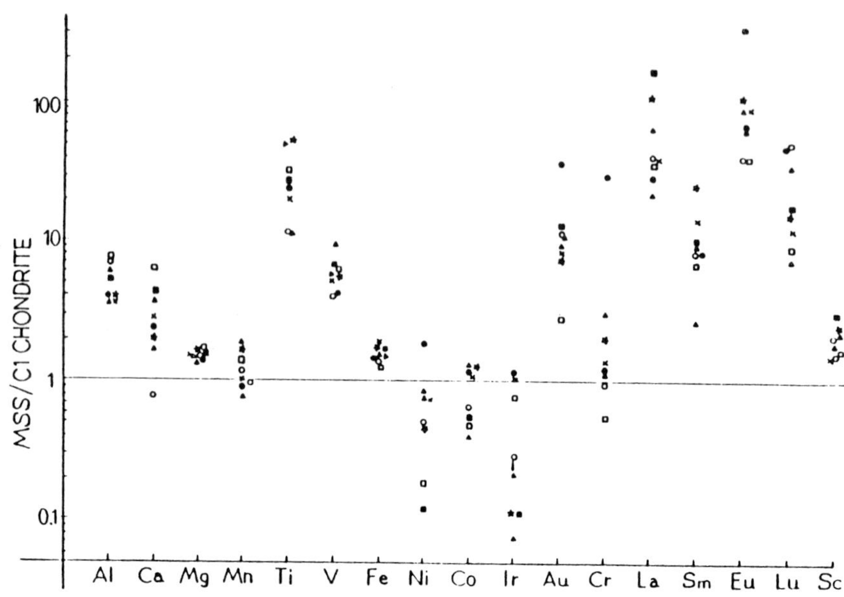

Fig. 7.3. The original compositions of stony spherules changed and partially oxidized, which are normalized with the solar abundance, are shown (Yamakoshi 1984).

Behaviors of elements in iron materials

Nogami (1985) studied the behavior of elements in iron meteorites. Small pieces of the "Henbury" iron meteorite and two additional kinds of alloys were melted in electric crucibles placed in a low (air) pressurized chamber.

The temperature was kept at 20°C above the melting point of each sample for 1 to 30 min and monitored with a pyrometer. The chemical compositions of the evaporation residues of the Henbury meteorite were determined by INAA for Fe, Co, Ni, Ga, As, Ir and Au, and those of Fe-Ni alloys were measured with EPMA for Si, Cr, Fe and Si.

In his experiments, a piece of the Henbury iron meteorite 27.6 g in weight was cut into smaller cubes of 1.0 mm in size. In order to check the homogeneity of the respective samples, seven of the cubes were analyzed by INAA and the total experimental uncertainty was estimated as not larger than ±5%. He determined the elemental compositions of the Henbury iron meteorite as [Fe] = 92.0%, Ni = 7.1%, Co = 0.46%, Ga = 18.3 ppm, As = 2.0 ppm, Ir = 11 ppm and Au = 0.59 ppm. The Alloys A and B are (Ni = 75.5%, Cr = 13%, Fe = 4.2%, Si = 4.2% and B = 2.8%) and (Fe = 92%, B = 3% and Si = 5%), respectively.

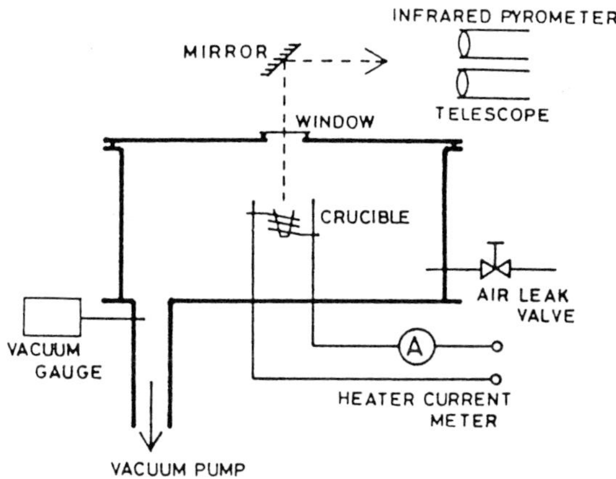

Fig. 7.4. The electric crucible used in the experiments is shown.

The melting experiments were carried out in a small electric crucible, made of boron nitride, placed in a vacuum chamber which was connected to a rotary pump. For higher vacuum tests, a diffusion pump was located before the rotary one. The crucible was heated from outside by a tungsten coil heater apparatus, and the temperature was monitored by an infrared pyrometer focused onto the crucible, and the heater current of the pyrometer was kept within ±5°C. The increasing rate of the temperature was 150°C/min. The melting features were observed by a telescope. Among the samples, a few were split and scattered at the melting temperatures and could not be recovered.

The melting temperatures were 1400 ± 5°C for the Henbury meteorite and 1000 ± 5°C and 1100 ± 5°C for Alloys A and B, respectively. In these cases, the definition of the melting points is difficult and indistinct, because the major materials such as Fe and Ni were melted while the impurity clusters and inclusions made of more refractory elements remained unmelted in the "liquid drops".

After the melting procedure, the remaining residue in the crucible was recovered and analyzed. The weights of the residues were compared with the original ones, so the ratio was defined as "residual fractions" (RF).

In Fig. 7.5 the behavior of the elements in the melting processes is shown. In order to compensate for the oxidization effect on the composition, the respective elements are shown here plotted as the ratios of the iron component.

In Fig. 7.5, we can see that the relative concentrations of Co, Ni and Ir are in inverse relation to the RF. However, Ga, As and Au decreased rapidly towards the detection limits for the respective elements by INAA.

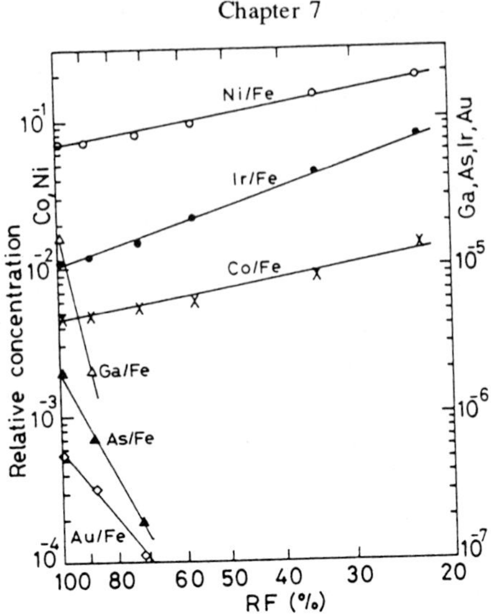

Fig. 7.5. The behaviors of the elements in Henbury (Nogami 1985).

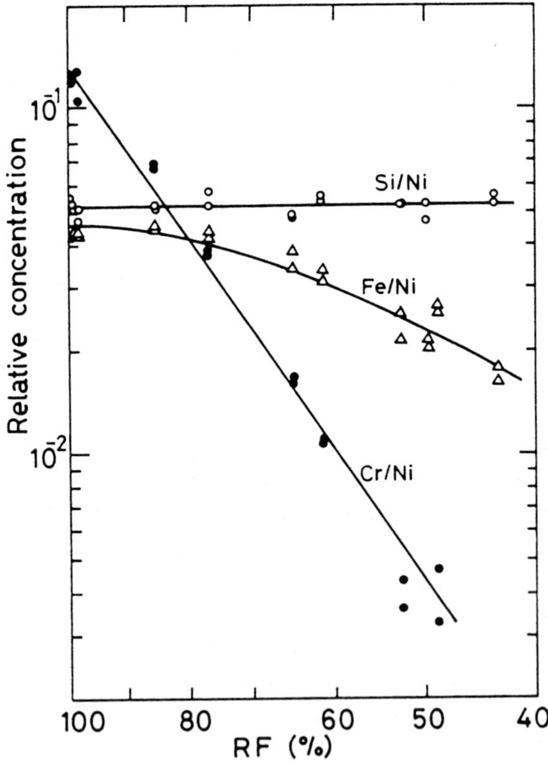

Fig. 7.6. The behavior of the elements in Alloy A (Nogami 1985).

Figure 7.6 shows the behavior of the elements in Alloy A. Relative concentrations of (Si/Ni) were kept as constants. In the nickel-based alloy, Si has the same evaporation rate as Ni, which is a very interesting feature for understanding meteoroid phenomena.

In these simulation experiments some interesting features of the elements were revealed. The chemical concentrations of the cosmic spherules were not analogous to ablated debris of iron meteorites. The conclusion can be drawn that the concentrations of the spherules before the melting processes are much different from those of iron meteorites.

7.2 Changes of Intensities of Induced Radio-Nuclides in Cosmic Meteoroids by Ablation Laboratory Experiments

For the purpose of estimating the degree of evaporation loss of cosmogenic nuclides from cosmic meteoroids in the ablation process by atmospheric heating, a few laboratory ablation experiments were performed by Nogami and Yamakoshi (1985). Small Fe and Ni metal chips in which radioactive ^{54}Mn, ^{51}Cr and ^{58}Co were induced by fast neutron irradiation were heated to 10~20°C above the melting point of iron under 40 m Torr air pressure. Gamma ray activities of the radionuclides measured before and after heating show that ^{54}Mn evaporates much faster than the parent material of Fe and that ^{51}Cr and ^{58}Co evaporate much more slowly than ^{54}Mn, but faster than the major elements Fe and Ni.

The origin of the magnetic iron spherules found in deep-sea sediments has been studied for the last two decades. We have adopted two criteria to identify the spherules as being of extraterrestrial origin. One is the detection of as many siderophile elements as in meteorites, which are enriched in "cosmic materials" (these materials are largely depleted in the Earth's crust, since the cosmic spherules are subjected to thermal degeneration during their atmospheric heatings; see Section 7.1). Another is the detection of cosmogenic nuclides in the spherules. However, it has been very difficult to identify these nuclides in individual spherules because of their small sizes. Recently, a new, highly sensitive technique known as AMS, or Accelerator Mass Spectrometry, has been developed using high energy particle accelerators with which it is hoped to be able to determine various cosmogenic, long-lived nuclides in individual spherules.

During atmospheric entry, cosmic materials are so ablated that some amounts of the cosmogenic long-lived, radio- and stable nuclides in these materials could be changed. This work (Nogami and Yamakoshi 1985) was done for the purpose of estimating the degree of change of the cosmogenic nuclides in the extraterrestrial materials. In order to do laboratory simulation experiments, some nuclides must be induced in the target material by artificial bombardment.

How can we simulate the process of the high energy cosmic ray irradiation in space for Aeons (10^8~10^9 yr) by short-term artificial bombardment?

In this work a few radionuclides were produced in the parent iron and nickel materials. ^{54}Mn and ^{51}Cr nuclides were induced in highly enriched ^{54}Fe metal chips through the reactions of ^{54}Fe (n, p) and ^{54}Fe (n, α) by fast neutron bombardments in a research reactor. ^{58}Co nuclides were also induced in highly purified Ni metal chips through a ^{58}Ni (n, p) reaction. After irradiation the samples were removed from the reactor, and evaporation experiments were performed.

Before and after the melting of these chips in a vacuum, the mass of the respective samples was weighed with a micro-balance, and the radioactivity was measured. The evaporation losses of these radionuclides were calculated for comparison with the residual masses of the parent materials.

Experimental procedures will be interpreted in details. Highly enriched ^{54}Fe metal chips (^{54}Fe = 99.65%, ^{56}Fe = 0.33% and ^{57}Fe = 0.02% in weight) and highly purified nickel metal chips (Ni = 99.97%, Fe < 15 ppm, Co < 0.5 ppm in weight) were packed in a cadmium capsule of 1.0 mm wall thickness to eliminate the thermal neutron flux by four orders of magnitude.

Neutron irradiation was performed for 40 hours in a research reactor, whose fast neutron flux was 5.5×10^{12} (n/cm^2, sec).

The heatings of the metal chips were performed in a small electric crucible made of boron nitride (BN), placed in a vacuum vessel. Small pieces of a few mg each in weight were put in the crucible and heated up to the melting point. After melting, the temperature of the crucible was kept 10~20°C higher than the melting point for several minutes. The temperature of the melt was measured by an infrared spot photometer and controlled by the heater current.

The vacuum condition was kept at about 40 m Torr of air pressure, which corresponds to 70~80 km altitude of the upper atmosphere and also the average altitude of meteor flashing (ionosphere; E-region).

Before and after the melting and evaporation in the vacuum, these irradiated metal chips were measured with a large volume Ge (Li) detector system. In the measurements of ^{54}Fe samples, small gamma ray peaks from ^{59}Fe were observed, which were induced by a small flux of thermal neutrons from trace amounts of ^{58}Fe contained in the "^{54}Fe samples" used in this work. However, peaks of 320 keV from ^{51}Cr and 835 keV from ^{54}Mn were clearly observed.

The amounts of ^{59}Co (n, 2n) in by-products of an impure Co component in the nickel samples could be estimated to be less than 10^{-8} of the ^{58}Co from the ^{58}Ni (n, p) reaction. In the measurements of the nickel samples, ^{60}Co activity from ^{59}Co (n, γ) was also negligibly small. The statistical errors of the radiation counting were better than ±1.0% for all cases; however, total experimental errors were estimated to be ±5%, mainly due to the reproducibility of sample mountings on the detector, weighing errors and also adsorption loss of small amounts of melted metals into the crucible.

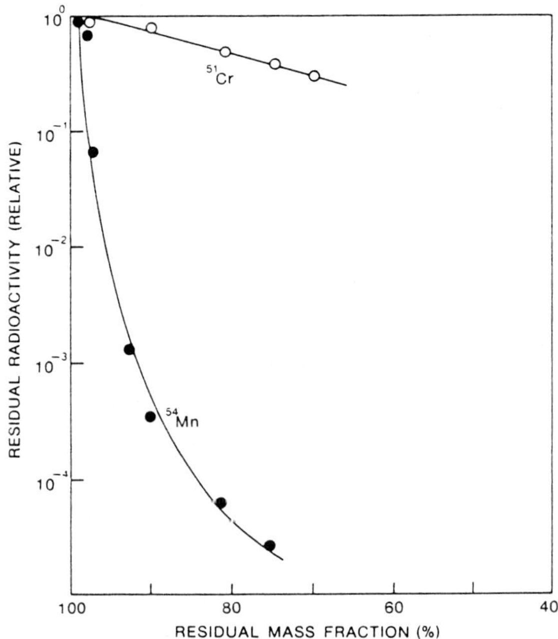

Fig. 7.7. Total ^{54}Mn and ^{51}Cr activities in the residues plotted against the residual mass fraction of the parent material of iron.

In Figs. 7.7 and 7.8, the measured radioactivity levels of the induced nuclides are plotted against the residual mass fractions of the parent materials. Here, the residual mass fraction is defined as a ratio of the weight of the residual materials recovered from the crucible after melting to their original weight.

^{54}Mn evaporates much faster than the parent material of iron. ^{51}Cr evaporates much more slowly than ^{54}Mn but faster than the parent material of iron.

^{58}Co evaporates in fairly good proportion to the residual mass fraction of Ni.

The behavior of ^{54}Mn in iron could be considered to be the same as that of the long-lived ^{53}Mn (3.7×10^6 yr), which is induced mainly through ^{56}Fe (p, α) reaction by solar proton irradiation in interplanetary space. Information on the behavior of ^{51}Cr in the ablation process of cosmic materials is also useful for studies of excess ^{53}Cr in primordial samples of the solar system.

The behavior of ^{58}Co also suggests that of ^{60}Co (5.271 yr) in nickel samples ablated by atmospheric heatings.

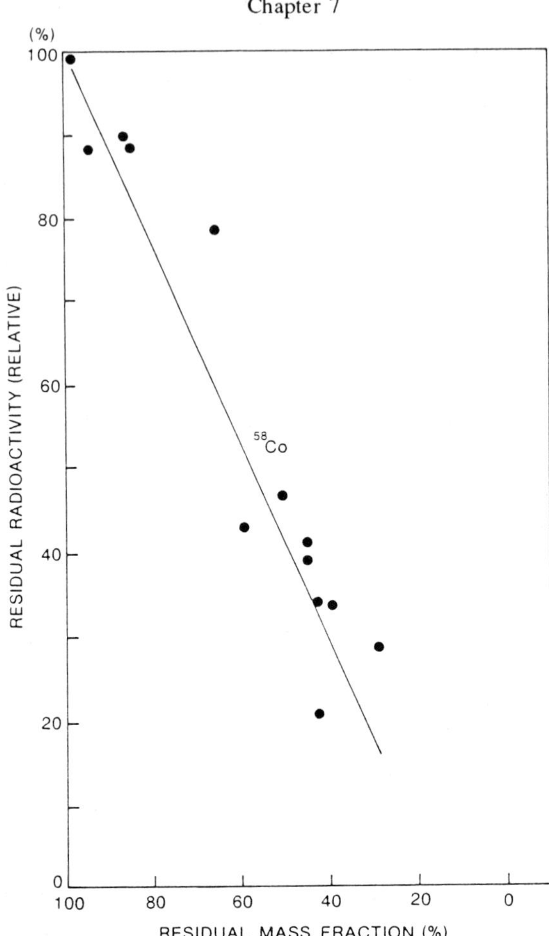

Fig. 7.8. Total ^{58}Co activity in the residues plotted against the residual mass fraction of the parent material of nickel.

Nishiizumi (1983) reported that ^{53}Mr activity was hardly detected in the deep-sea iron spherules with one exception. This suggests that most iron spherules had lost more than 95% of their pre-atmospheric weight.

The exception, which had considerably high ^{53}Mn activity in Nishizumi's work, might have been ablated so moderately during its atmospheric entry that more than 98% of the initial mass remained.

In this work (Nogami and Yamakoshi 1985), only three radionuclides could be produced by fast neutron irradiation for the laboratory experiments. Among the long-lived cosmogenic nuclides, which are expected to give valuable information on the origin and trajectory of interplanetary dust, only the behavior of ^{53}Mn could be simulated.

7.3 Changes of Elemental, Mineralogical and Textural Compositions by Laboratory Simulation Experiments

Blanchard *et al.* (1978) performed simulated ablation experiments for iron and iron-nickel samples, which contained traces of Mn, Cr, V and Si as impurities. Their experimental conditions were prepared to simulate meteor velocity of 12 (km/sec) and 70 km in altitude using an arc-heated plasma of ionized air, argon gas and electrons. The duration time was chosen as 30 sec for each.

In these studies natural fusion crusts of three iron meteorites, Boguslavka (II A), Norfork (III A) and N'kandhala (III A), as well as eight magnetic spherules from ferro-manganese nodules were used and compared with each other.

During the ablation process, droplets of melted alloy fell down, while others formed flanges on the flanks of the samples.

A thin (50~80 μm in thickness) fusion crust formed, which was composed mostly of an intergrowth of magnetite with an amorphous glassy phase. The amorphous glassy phase formed along the outer surface of the bulk samples. Several magnetite grains were observed to have Cr-rich cores. Adjacent to the fusion crust was a heat-affected transition zone (ca. 200 μm in thickness) in which spherical oxide particles were formed from the melted phase.

Table 7.1. The chemical compositions of the fused crusts before and after the ablation experiments using microprobe measurements (Blanchard *et al.* 1978).

Element	Iron sample		Iron-Nickel sample	
	Unablated	Ablated	Unablated	Ablated
Fe	99.1	99.8	89.9	90.8
Ni	0.08	0.08	9.7	9.9
Si	0.15	0.05	0.40	not detected
Mn	0.10	0.01	0.15	0.02
Cr	0.10	0.08	0.07	0.07
V	0.02	0.02	0.01	0.01
S	trace	not detected	trace	not detected
P	trace	not detected	trace	not detected

Table 7.1 shows the chemical compositions of the newly formed phases in the fusion crusts of iron meteorites, in which core-formation was observed in several magnetic grains. Blanchard and his co-workers noted that more than 90% of ablated particle grains were aggregates of spherules or single spherules. Ablated particles were typically divided into two types. The most abundant type is a grain, whose metallic core is surrounded by an iron oxide shell, and the other type is glassy material which contained many small metal or iron oxide inclusions.

The most common "mineral association" found in individual spherules is that of an α-iron metal core surrounded by a shell which consists of magnetite and wuestite.

Hematite was also found as a primary mineral in ablation debris; however, it depended on the level of oxygen during melting. The glassy spherules, which contained 10% or less of all ablation debris, were typically amorphous and yielded weak-α-iron or magnetic patterns only when they contained many metallic inclusions.

The natural fusion crusts were compared with artificially produced ones. The natural fusion crusts contain discontinuous vesicular iron oxide shells from 50 to 300 μm or thicker. Adjacent to this zone, there is a heat-affected metallic transition zone of several mm or less, which no longer contains structures, such as Widmannstaetten bands, troidal inclusions, or schreibersite lamellae.

Table 7.2. Chemical compositions of new phases formed in the fusion crusts of natural iron meteorites.

			(% by weight)
Element	Magnetite	Cr-rich phase	Dark-gray glassy phase
Fe	70.5 ± 2.0	30.9 ± 2.0	50.6 ± 4.0
Mn	1.0 ± 0.6	2.2 ± 0.5	5.0 ± 1.2
Cr	0.06 ± 0.05	33.8 ± 2.0	0.04 ± 0.02
V	0.04 ± 0.02	3.5 ± 1.0	0.03 ± 0.02
Si	0.15 ± 0.05	not detected	10.9 ± 1.5

Usually, Widmannstaetten bands are composed of alternating band layers of kamacite and taenite; however, both components transformed to the inequilibrated α_2Ni-Fe form by heating and therefore by rapid cooling. Troilite and schreibersite, $(Fe, Ni)_3P$, were completely melted by heating, leaving large cavities in the fusion crusts. They might be recrystallized to fine grained aggregates in the upper atmosphere.

Hematite was contained in natural fusion crusts as circular inclusions, 0.5 μm or less in size.

In the conclusion of the studies, Blanchard and his co-workers stated that the mineral compositions of the natural fusion crusts of meteorites are very similar to those of the fusion crusts artificially produced by the ablation process as well as magnetic spherules recovered from deep-sea ferromanganese nodules. They also said that melting at high temperature and rapid cooling produced melt features and preserved disequilibrium mineral phases such as α_2-iron and wuestite.

Volatile elements such as S, P, and Cl were vaporized. Some of these elements might be recrystallized in the upper atmosphere. Refractory elements were offered fractionations and concentrated into different phases according to their affinity for oxygen.

Therefore, the spherules were formed in the early stage of ablation and when the degree of oxidization was not so large, since Si and Mn components were concentrated into spherules produced in the early stage of ablation. However, a high nickel concentration was observed in natural spherules, which might be ablated from octahedrite (melted taenite) or nickel-rich ataxite in parent meteorites. Fusion crusts of natural meteorites were formed at the later stage of ablation in the atmosphere.

Blanchard *et al.* (1980) discussed the actual spherules collected from deep-sea sediments. They stated in the paper that they extracted spherules magnetically from deep-sea sediments and divided them using dominant mineralogy into three groups: iron, glassy and silicate.

They found relict grains in silicate spherules, such as olivine crystals, enstatites, ferrous spinels, chromites and pentlandites. Studies of fusion crusts of natural iron meteorites and ablation experiments on silicate minerals have informed us that separated metal spherules are never produced from ablated, predominantly silicate meteoriods.

The glassy spherules are more iron-rich than the silicate spherules. They consist of magnetite and an iron-rich glass and are relatively low in the silicone component. The silicate spherules undoubtedly originated from ablated stony meteoroids. Olivine-magnetite-glass and sulfide phases have been newly originated.

Relict grains are used as a good clue to search for parent bodies before the ablation process in the upper atmosphere.

On the basis of experimental results, Blanchard *et al.* (1980) proposed the criteria useful for recognizing ablated extraterrestrial materials:

(1) Disequilibrium mineral phases are found, such as wuestite, glass and zoned olivine, due to melting and rapid cooling.

(2) There are distinctive textures, such as zoned olivine surrounded by magnetite, which is produced by intergrown dissimilar phases.

(3) Depletion and migration of volatile elements due to atmospheric heating, such as S, P, C, Cl, Na and H_2O.

(4) In some mineral phases chemical fractionations and also concentrations of less volatile elements due to their affinity for oxygen occur, such as Ni-rich metal cores surrounded by Ni-poor oxides.

In the case of iron parent meteoroids, the metal fraction is oxidized to produce wuestite, magnetites and hematites. Since nickel is not as easily oxidized as iron, the Ni component concentrates into a metallic core. When magnetite is not produced by ablation, the core has a very low iron content.

In the case of chondritic parent bodies, the silicate metal phase is dominant and crystallizes to form grains of zoned euhedral olivine and magnetites in a Ca, Fe, Al, and Si-rich glass matrix. The other phase contains Fe, Ni and S and crystallizes to form iron sulfide and iron oxides.

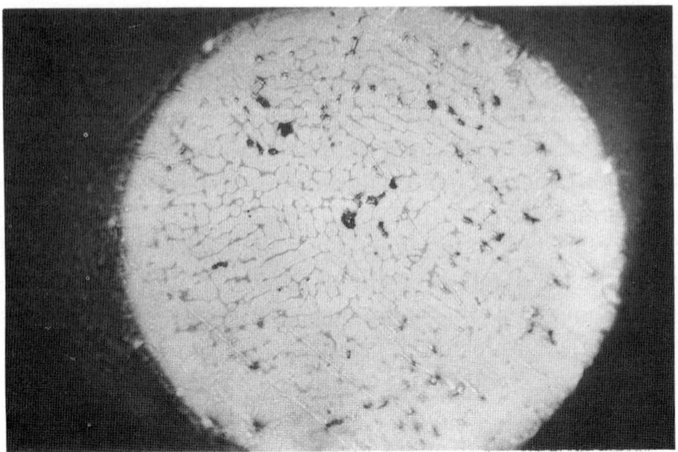

Fig. 7.9. SEM picture of a cross-section of an iron spherule with intergrowths of magnetite (dark) and wuestile (light) without core structure (photo by Dr. K. Nogami).

7.4 *Changes of Chemical and Isotopic Compositions of Ejecta Induced by Hypervelocity Impacts*

In order to study changes in the chemical compositions of ejecta scattered from dust-dust and dust-meteorite collisions in space, two laboratory hypervelocity impact experiments by two-stage light gas gun were performed (Yamakoshi *et al.* 1985). In this work, we measured only changes in the chemical compositions of ejecta; we will perform some experiments to study the changes of isotopic compositions in ejecta in the near future. Changes in chemical composition including noble gas contents as well as mineral phases in shocked chondrites and iron meteorites have been studied from the point of view of collision-induced metamorphism of cosmic matter in space.

As a common material, the stainless steel SUS 304 was chosen for both projectiles and targets, because SUS 304 steel is an industrial product of which we can expect a good physical and chemical homogeneity, and it was also thought to be resistant to oxidization in the impact processes, as suggested by the work of E. Gault *et al.* (1972).

In order to avoid oxidization of the materials during the course of impact, the impact chamber was kept in a vacuum (~0.1 Torr). However, this vacuum condition was not adequate to protect the smallest grains of the ejecta against oxidization.

A large amount of styrofoam blocks and magazine papers which were packed in the impact chamber caught the ejected fragments without much deformation and loss.

The chemical composition of the material was similar to those of iron meteorites and/or the metal phases in ordinary chondrites. They were analyzed using INAA; the homogeneity of the raw material (SUS 304) used in this work was also checked with an X-ray microanalyzer. The projectiles were fired by a two stage light gas gun, which can accelerate a bullet of ~1 g up to 6 km/sec.

In the first run, a cylindrical projectile (4 mm diameter × 2.15 mm length) of mass 0.209 g which was mounted in front of a polyethylene sabot (7 mm diameter × 10 mm length) of mass 0.350 g was impacted normally against the facial center of a cylindrical target (6 mm × 2.7 mm length) of mass 0.612 g. The velocity of the projectile was 3.6 km/sec. The energy density (total kinetic energy of the projectile/mass of the target) was 2.21×10^{10} (erg/g).

In the second run, a cylindrical projectile (4 mm diameter × 9 mm length) of mass 0.152 g, which was mounted on a nylon sabot (7 mm diameter × 1.6 mm length) of 0.304 g, struck normally a cubic target (50 × 50 mm, 30 mm length) of mass 594 g. The velocity of the projectile was 3.86 km/sec. The crater produced was 16 mm in diameter (inside rim) and 4.5 mm in depth (excluding rim height). The energy density was 1.9×10^7 (erg/g). The 0.1 Torr of air pressure was maintained in the impact chamber by a rotary pump.

The chemical compositions of the ejecta and the raw material were determined by INAA with a TRIGA II reactor. The raw material (SUS 304) was checked with a reference sample (Canyon Diablo) of homogeneous form, and Cr and Sb were obtained with JB-1 glass reference. We determined the chemical composition of the SUS 304 as [Fe] = 75.6%, [Ni] = 6.35%, [Co] = 0.152%, [Cr] = 17.7% and [Sb] = 0.33 ppm in weight. The inhomogeneity of chemical composition in the raw material was checked in seven chips cut from the raw material, whose data coincided within ±5%. SUS 304 was also checked by X-ray microanalysis; the ratios of K-X ray intensities of Cr to Fe and Ni to Fe were obtained from over 40 spots on the surfaces of SUS 304 (each spot size, 40 × 300 μm) as 0.4283 ± 0.0131 for (Cr/Fe) and 0.104 ± 0.00252 for (Ni/Fe).

Run 1; Grain to grain collisions

In Run 1, a catastrophic fragmentation of the target and the projectile occurred. The scattered fragments were gathered from the styrofoam and magazine papers packed in the impact chamber. The styrofoam was dissolved with toluene and the lodged ejecta were collected. The small fragments lodged into the magazine papers were carefully removed by turning over the pages. However, the total weight of the scattered fragments recovered from the packing material amounted to at most only 20% of the total amount of the target and projectile.

Therefore, a size distribution of the scattered debris could not be obtained. No spherical grains were found among the collected debris.

The chemical compositions of the collected fragments were determined by INAA.

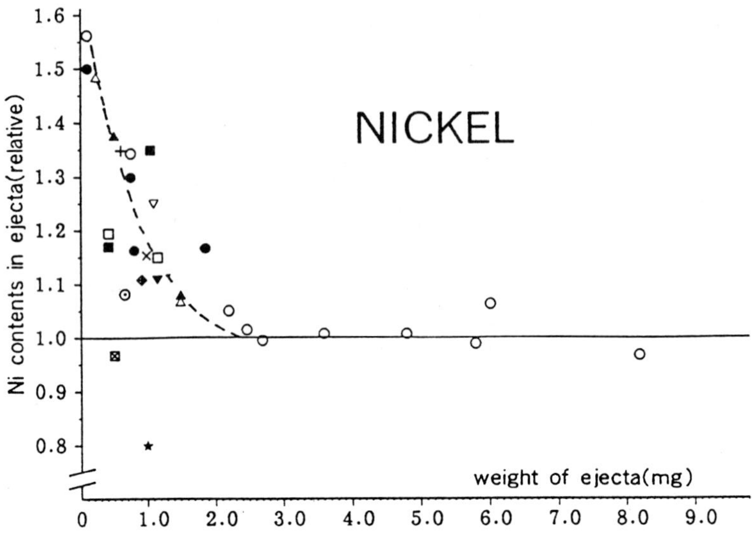

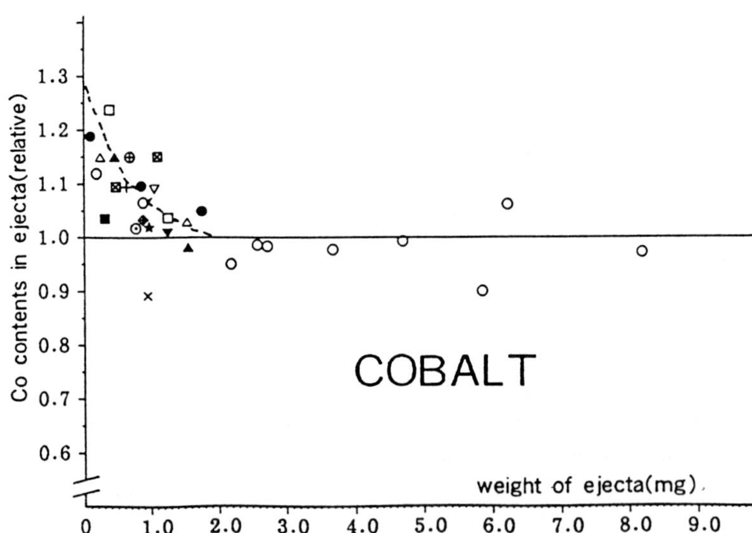

Fig. 7.10. Ni (upper) and Co (lower) contents in the fragments of ejecta, relative to those of the raw material, obtained in RUN 1. Various symbols are used for identification of the respective samples in the figure.

In Fig. 7.10, Ni and Co contents in the grains smaller than 3 mg become more concentrated as the grain mass decreases. In contrast, Sb contents are seen to be depleted in inverse proportion to the grain mass, which is shown in Fig. 7.11.

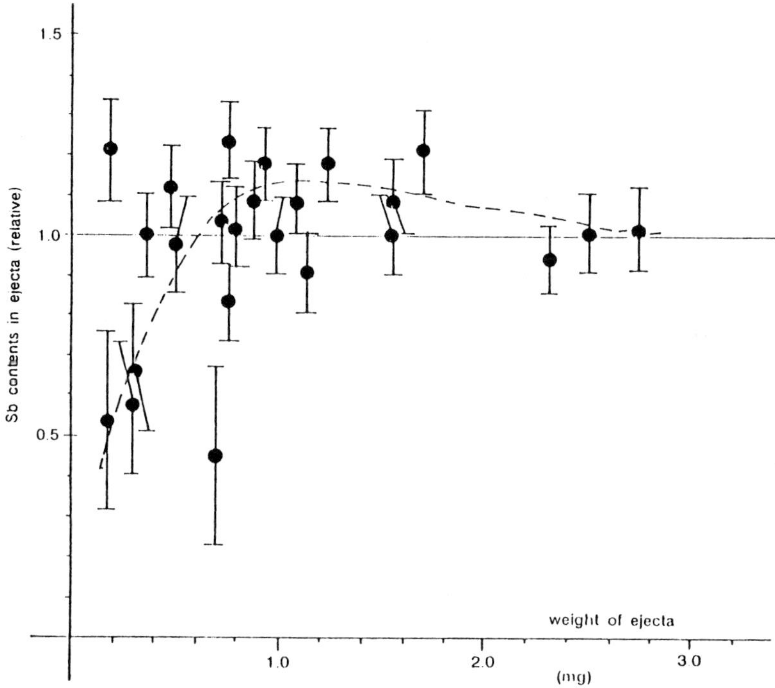

Fig. 7.11. Sb contents in the ejecta fragments, normalized with Sb value of the raw material (SUS 304).

In Fig. 7.12, the behaviors of Cr and Fe contents are shown. The Cr contents are nearly constant through the entire mass region except for a few cases. Fe contents are widely scattered; however, in general Fe is depleted in all cases.

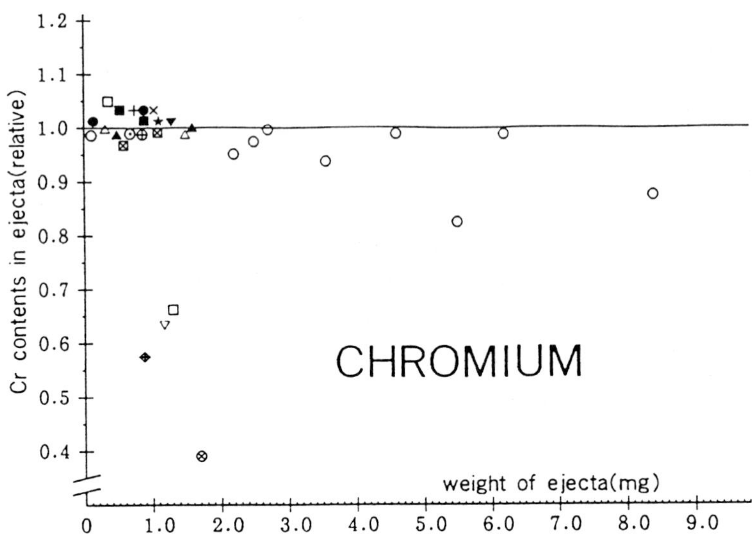

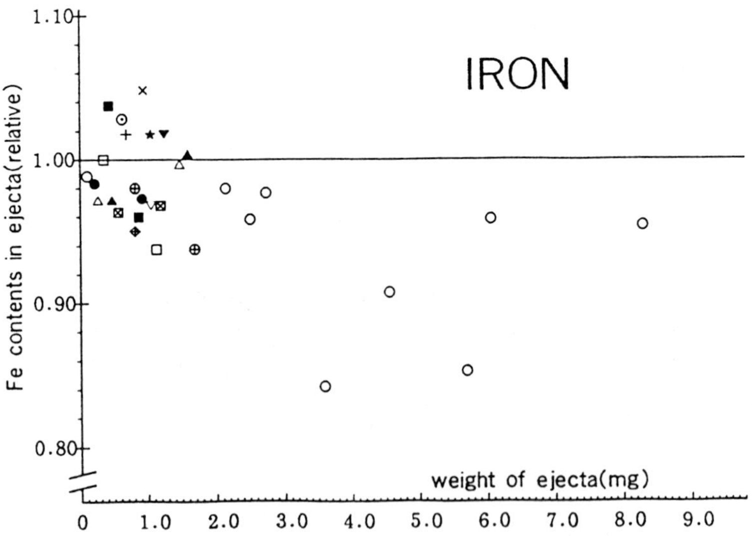

Fig. 7.12. Cr and Fe contents in ejecta (Yamakoshi *et al.* 1985).

Run 2; Grain-to-large-body collisions

The [diameter/depth] ratio of the crater left on the target was 3.6, and the mass ratio of the ejecta to the projectile was 2.7. No spherical grains were found in the collected ejecta in this run. Many fragments had twisted shapes, on which a special pattern just like a rifle mark was observed. This pattern originated from the target surface, which was processed by a milling cutter. Some of the fragments looked burned up and had changed to a blue-violet colour. Over 90% of the ejected mass was recovered, which was the amount of the projectile plus the ejected mass from the target crater.

The mass distribution of the ejecta is shown in Fig. 7.13. The distribution of elemental compositions of the smaller ejecta was also determined by INAA.

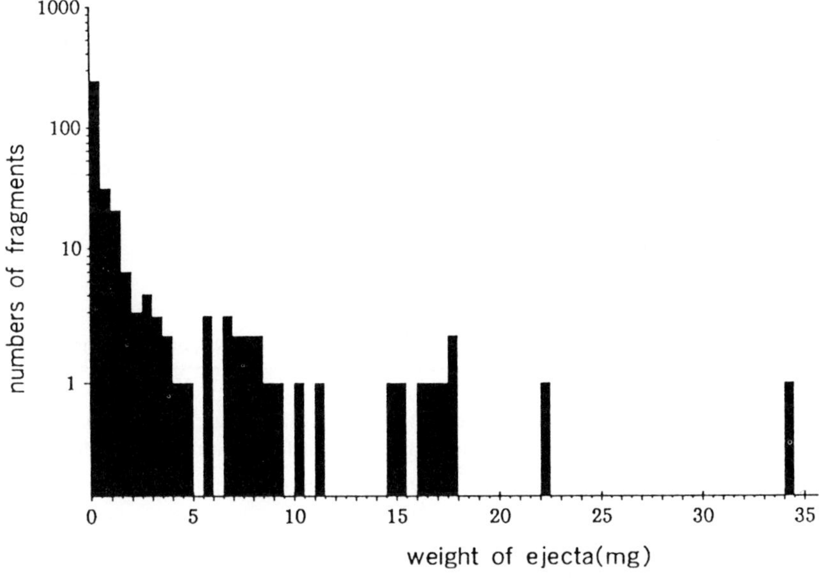

Fig. 7.13. Mass distribution of the ejecta in RUN 2 (Yamakoshi *et al.* 1985).

The chemical compositions of the ejecta are shown in Fig. 7.14. The contents of Fe, Ni, Co and Cr in the ejecta are normalized with those of the raw material.

In Run 1, the target was so small-scaled that it was broken and scattered around, while in Run 2 only cratering occurred in the target block. This could be explained by the fact that the former energy density was much higher than the latter by three orders of magnitude (Fujiwara *et al.* 1980).

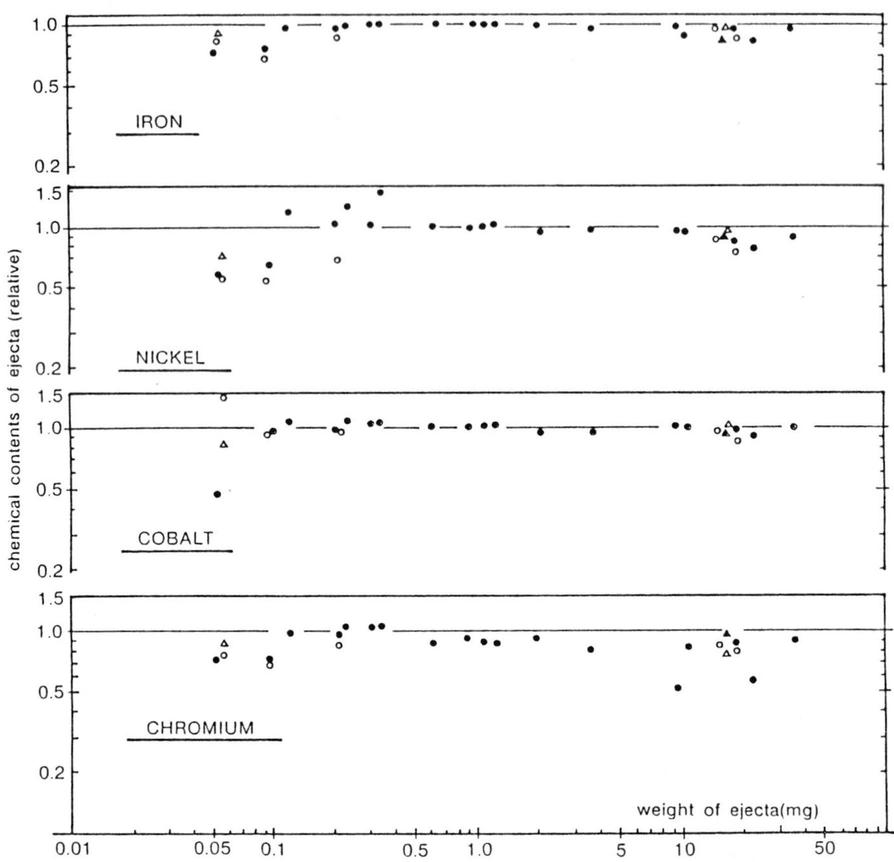

Fig 7.14. Fe (upper), Ni, Co and Cr (lower) contents in the ejecta fragments, normalized with those of the raw material. Various symbols are used for identification of the respective data in the figures (Yamakoshi *et al.* 1985).

The Rankine-Hugoniot one-dimensional calculation gives the originated pressure (P) and the peak temperature (T) at the shock front of the impact:

Run 1; P = 1.02 Mbars, T = 1600 K,

and Run 2; P = 1.12 Mbars, T = 1850 K.

In these experiments, the reaction time duration is indeed very short; however, the given pressure is much higher (just three orders of magnitude) than that produced by hydrostatic pressurizing techniques. The shock loading duration times were estimated as 0.67 μsec for Run 1 and 0.69 μsec for Run 2.

In these experiments the enrichment of Ni and Co components occurred largely in smaller fragments, which might have been the result of the strong shock, and were restricted to the small volume adjacent to the impact site. Despite the poor radiation counting statistics, depletion of Sb contents also occurred in the smaller fragments in Run 2. In these fragments we could find an inverse correlation between Ni and Sb, in which the Co contents were also enriched.

Any distinct causal and sophisticated relation could not be obtained here. However, the enrichments of Ni and Co might be caused by the depletion of the most abundant component, Fe, in the ejecta. In the behavior of Cr contents in the ejecta, no distinct size dependence was seen in either experiment.

The results obtained here suggest the possibilities of the enrichment or depletion of chemical components in ejecta during dust-dust and dust-meteorite impact phenomena in space.

Meteorite-dust collisions are occurring in the present time in space, thereby we can expect that the compositions of the ejected grains will be changed more than those of meteorites. Dust-dust collisions might have occurred in highly dense dust-clouds at the time of birth of the solar system. Therefore, studies of changes of chemical and isotopic compositions as well as sticking probabilities of dust-dust collisions are important subjects in the cosmosciences.

In this work (Yamakoshi *et al.* 1985) we could not examine the changes of isotopic fractionations due to the hypervelocity impact phenomena.

However, in order to understand exactly the enrichment and depletion of the elemental compositions in the ejecta, further impact examinations of various materials are necessary, because our results were obtained only with regard to some special sets of experimental conditions: materials, target sizes, velocities of the projectiles and vacuum conditions (Yamakoshi *et al.* 1985).

Many problems remain unsolved: what changes will occur in cases caused by hypervelocity impacts; cosmogenic nuclides in cosmic targets and Eu or Yb anomalies in REE patterns in cosmic matter, etc.

REFERENCES

Blanchard M. B. and Cunningham G. G. 1974, J. Geophys. Res. 79 3973.

Blanchard M. B. and Davis A. S., 1978, J. Geophys. Res. 83 B4 1973.

Blanchard M. B., Brownlee D. E., Bunch T. E., Hodge P. W. and Kyte F. T., 1980 Earth Planet. Sci. Lett. 46 178.

Fujiwara A 1980, ICARUS 41 356.

Gault E., 1972, private communication.

Nishiizumi K., 1983 Earth Planet. Sci. Lett. 63 223.

Nogami K. and Yamakoshi K., 1985, Geochem. J. 19 97.

Nogami K. 1985., Geochem. J. 19 101.

Notsu K., Onuma N. Nishida N. and Nagasawa H., 1978, Geochim. Cosmochim. Acta 42 267.

Shepard C. E., 1967 NASA Tech. Note D-3740.

Yamakoshi K., 1984 Geochem. J. 18 147.

Yamakoshi K., Fujiwara A . and Asada N., 1985 J. Geomagne. Geoelectr. 37 987.

Chapter 8

In-Situ Measurements of Interplanetary Dust

8.1 *Various Detection Methods and Detection Devices*

Since the interplanetary dust is so fine and delicate, its physical and chemical properties would be better analyzed in-situ by spacecraft-borne detectors. In this section, a research history will be followed, and several analyzers and obtained results are reviewed.

In the 1950's in-situ dust detection was tried by rocket-borne and satellite-borne detectors, whose main detection function was designed with piezo-materials. Many American and Russian scientists stated that their space-borne detector "discovered" dust-belts around the Earth.

However, during the next decade it was found that the "impact signals" had originated as piezo-noise from structural disturbances of the piezo materials due to sudden temperature changes in the upper atmosphere. Because of the piezo-detector's problems, rapid developments of other detection methods were required; one candidate was "pad-sensors", which were pressurized "pads" with inert gas. When a dust collides with the pressurized pads in space, the pads are broken, loosing the inert gas, and installed transducers emit the impact signals.

The broken pads cannot regenerate again, so this detection system is often called "not healthy".

However, as they were only punctured by dust impacts, no further information, such as velocity, incoming angle or mass, of the dust was obtained. Pad-networks were borne on Pioneer #8 and #11 and brought us fruitful information on the spatial influx of the interplanetary dust in the regions of the Asteroids, Jupiter and Saturn (Humes *et al.* 1975).

Two-frame network sensors, whose network frames are tightly covered by thin films, such as aluminized mylar films and piezo-films (PVDF = polyvinylidene fluoride; Simpson *et al.* 1985), are also effective detectors, and are able to determine the incoming angle by penetration at (X_1, Y_1) of a front panel and (X_2, Y_2) of a rear panel, and the velocity by the time-of-flight method using the time difference between the front and rear panel signals. If a thick stopping plate is prepared at the bottom of the array, the total impulse lost in the plate informs a complex product~$m \times v$, where m is a mass and v is a velocity of an impact grain.

A concept of a two-frame network detector is shown in Fig. 8.1. Thin piezo films tightly cover the frames, and the impact positions can be determined by IC network signal treatments.

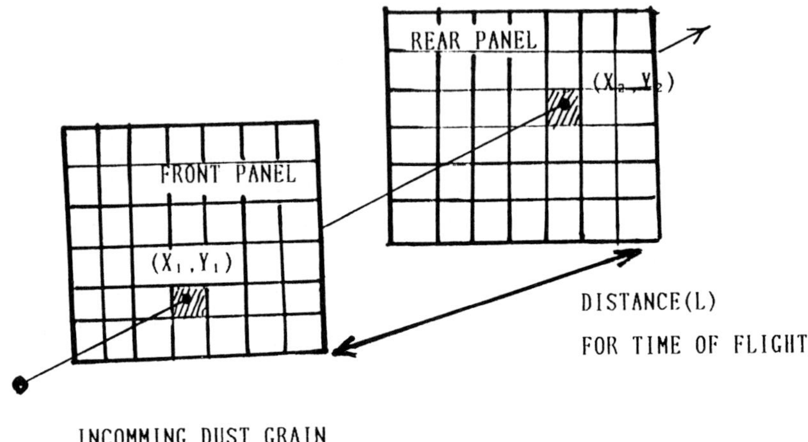

Fig. 8.1. A schematic diagram of the two-frame network detector system.

Iglsder and Igenbergs (1987) and Igenbergs *et al.* (1991) prepared and mounted a Faraday-Cup device onto a Japanese engineering satellite"MUSES-A", launched in January, 1990. The function of the device is shown in Fig. 8.2.

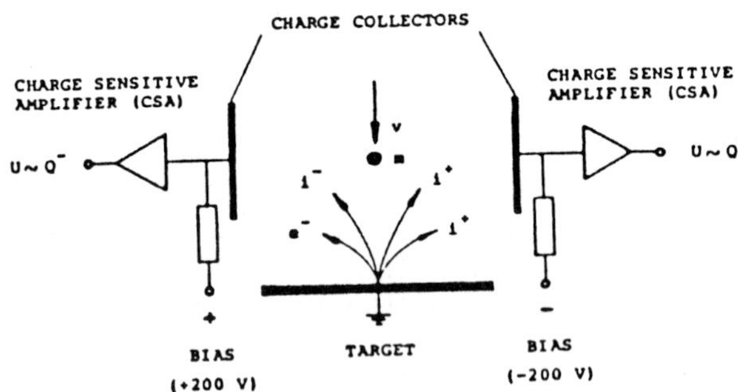

Fig. 8.2. An Impact Ionization Device borne on MUSES-A (Iglseder and Igenbergs 1988).

With this device they are planning to measure the total charges generated by dust impacts, and an empirical formula is expressed,

$$Q(\text{total}) = C \cdot M^{\alpha} \cdot V^{\beta} \quad \text{(Coulomb)},$$

where C is a function of the density ratio of (target/projectile), α is approximately unity and β is a value between 2.92 and 3.77.

As discussed in Section 9.2, (D/T) ratios of microcraters formed by direct collisions of interplanetary dust onto meteoroids, such as lunar rocks or large meteorites, are dependent on the material combination of the target and the projectile but are independent of the velocity of the impact particles. Yamakoshi (1982) proposed a design so that as a dust grain hits on a surface plate of a satellite, whose points of impact are detected by scanning the impact signals, the depth and diameter of the microcrater are automatically measured with light sensors. Then we can know which type of projectiles hit the plate, stony or iron?

Impact-plasma with TOF (Time Of Flight method) is also a most fruitful and promising detection method for the elemental and isotopic compositions of the colliding dust grains (Gruen and Fechtig 1978).

The plasma generated at the instant of the impact at the window of the detector is accelerated by a few electrodes and moves a linear distance to the stopping plate. The time difference of the starting (trigger) and stopping signals determines the mass of the ionized atom. This method is effective, of course, for the simplest mass spectrometry in laboratories.

An impact plasma detector, carefully designed and improved, could gather also various excellent data about P/Halley's cometary grains. The fundamental structure is shown in Fig. 8.3 (Fechtig, Gruen and Kissel 1978).

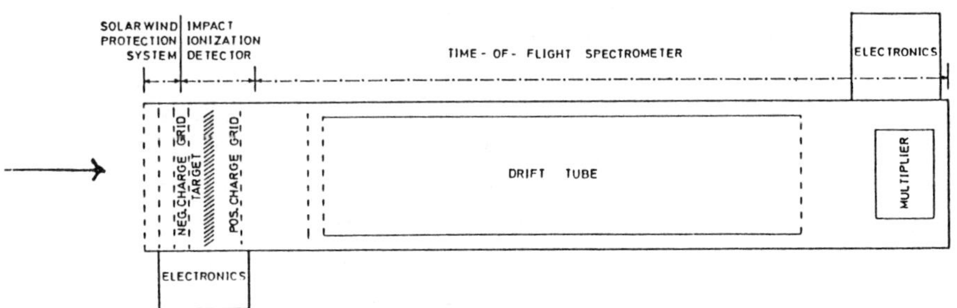

Fig 8.3. The schematic structure of an impact plasma detector (Fechtig, Gruen and Kissel 1978).

The longer the ion path length becomes, the better the resolution of elements and isotopes which can be obtained. Therefore, some variations of impact plasma detectors are proposed in which the ions take longer paths by folding back in the long drift tube and supplying electric fields to turn over the ions.

Here an exercise calculation is performed: a fine grain hits the window of the analyzer in a vacuum and changes to a plasma cloud with its own large kinetic energy, whose ions are mainly single-charged.

The initial velocity of the ions generated there is assumed to be nearly zero, and the applied voltage between the drift tube (length: L) is E. The time durations in which generated heavy ions fly the length of the drift tube (L) are signed as t_c for ^{14}C and t_f for ^{56}Fe. So we have

$$M \cdot a = q(E / L) \tag{1}$$

$$L = 1/2 \cdot a \cdot t^2 \tag{2}$$

where M is a mass of ions, a is an acceleration constant, and q is a unit charge. From equations (1) and (2),

$$t = L \cdot \sqrt{2M / qE} \tag{3}$$

Here we put into equation (3):

L = 1 (m), E = 5000 (volts), q = + 1e = 1.602×10^{-19} (Coul.) and M_c = 2.325×10^{-26} kg for ^{14}C and $M_f = 9.299 \times 10^{-26}$ (kg) for ^{56}Fe, so we have t_c = 7.62×10^{-6} sec, $t_f = 15.24 \times 10^{-6}$ sec, and the time difference $\Delta t = t_f - t_c = 7.62$ μsec. Thereby we can obtain simply a flight time spectrum, i.e. a mass spectrum of the plasma cloud which arose from the impact dust.

In the near future, plasma generated at the instant of the dust collision can be analyzed precisely, and spectra of the chemical and isotopic compositions of the projectiles will be taken. A simple detector has been proposed using large area solar cell panels of the spacecrafts (Yamakoshi 1988). Spacecrafts aimed to launch toward the Sun region always have a solar cell power network. If we can complete a simple dust counter using solar cell panels, we will have many chances to mount the dust counter on board every spacecraft flight. In this cell counter, a DC power channel is supported to generate solar power for the housekeeping, scientific works and telecommunications. However, a dust impact pulse will be lead to a pulse channel through a condensor route.

If we wish to improve the angle resolution of the incoming dust on the solar panel, one of the good answers is to make a barrier-network on the solar cell paddles. Each cell is surrounded by rectangular barriers which are lined with shining metals; they become mirrors to reduce no sunlight power for the DC channel. If we choose the barrier height to be the same size as those of the solar cell dimension, we have a certain angle resolution.

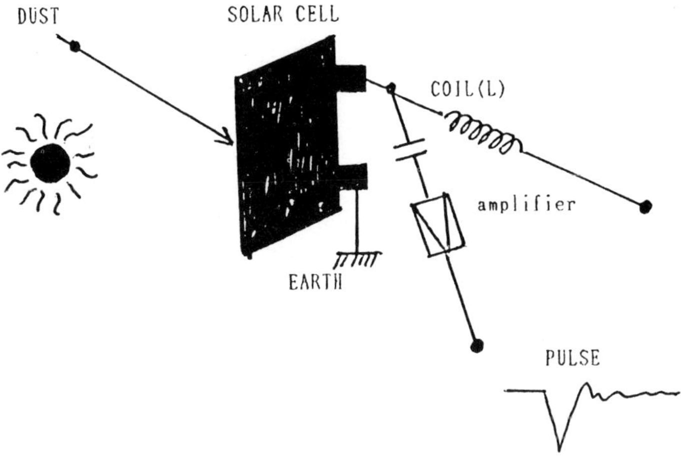

Fig. 8.4. A schematic diagram of a simple dust counter using a solar cell (Yamakoshi 1988).

8.2 Some Dust Catcher Designs with Lining Materials of Microcraters

At the time of hypervelocity impact collisions of dust with a target, it is expected that some residue materials of the projectile may remain inside the micro-crater, just like impactites, such as impact glass or coacites, can be found in meteorite craters on the Earth. Hallgren and Hemmenway (1976) detected the elements in Table 8.1 from the target materials of the SKYLAB mission.

Table 8.1. The detected elemental compositions of the impact residues found in the SKYLAB mission (Hallgren and Hemmenway 1976).

Type		Crater Diameter (μm)	Crater's [Diameter/Depth]	Detected Element
Smooth	B-1-12-2	5.2	4.3	Al
	C-2-0-0-2	1.8	—	Cr, Fe, Ni
Melted	B-1-12-1	47	2.1	Fe
	A-3-15-2	21	1.8	Al
	A-3-16-2	20	1.9	Al
	A-3-16-3	31	1.9	Al
Lumpy	B-1-12-4	15	3.9	Mg, Al, Si, S, K, Fe, Ca
	B-2-9-3-1	1.8	—	Si, S, Zn
	B-3-15-9-1	12	2.6	Al
Tex-tured	B-1-12-3	7.8	2.8	Mg, Si
	D-2-16-8-1	2.5	—	Al, Si
	A-3-15-1	7.8	3.0	Al
	B-3-15-1	6.9	2.5	Si, S, Fe
	C-3-14-1	5.6	3.1	Al, Si, S, Fe
	C-3-14-2	5.4	3.6	Al
	C-3-14-4	4.9	4.1	Al
	C-3-14-5	6.5	3.3	Al
	C-3-16-1	2.1	—	Al, Si

Gaswerk *et al.* (1982) measured residues of the projectiles which were made of iron meteoroids and estimated the total amounts of the residues in the craters, which are shown in Table 8.2.

Table 8.2. Estimated projectile's residues in microcraters in simulation experiments.

Target material	Projectile velocity (km/sec)	Crater size (mm)	Measured element	Residual projectile material (mg)	(%)	Analytical technique (*)
Au	4.1	6.7	Fe	9.15	57	AAS
W	3.9	3.5	Fe	0.48	3	AAS
Al	4.5	6.0	Fe	0.32	2	AAS
Glass	4.1	3.0	Fe	16×10^{-3}	0.1	AAS
Glass	4.0	2.9	Fe	8×10^{-3}	0.05	NAA
Glass	4.1	3.0	Fe	6.4×10^{-3}	0.04	NAA
Glass	4.1	3.0	Cr	6.0×10^{-4}	0.23	NAA

*AAS = Atomic Absorption Spectrophotometry, NAA = Neutron Activation Analysis.

In the following figures I will demonstrate the features of the residues of projectile materials inside the microcraters in our experiments.

In our case we collided a tiny iron-alloy particle (Fe = 94%, Ni = 6%) against thick aluminum targets. Figure 8.5 shows a cross-section of the #1 crater where the diameter was 15 mm and depth was 16 mm; therefore, the ratio (D/T) was 0.94.

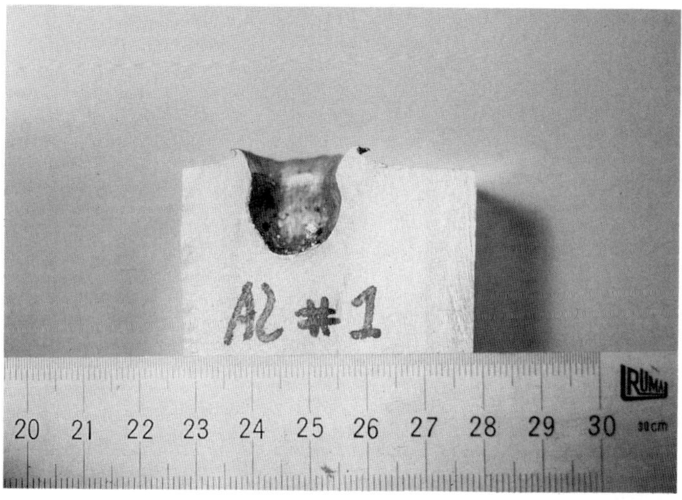

Fig. 8.5. A cross-section of a micro-crater. Target is Al, and the projectile is Fe-Ni alloy (Fe; 94%, Ni; 6%).

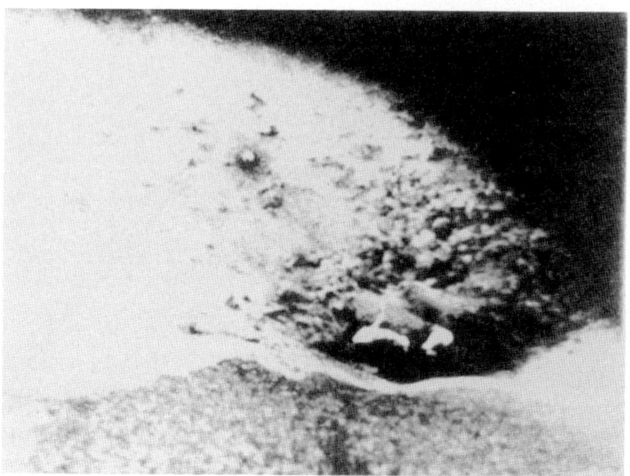

Fig. 8.6. A small portion of the bottom of a micro-crater taken with a SEM.

Figure 8.6 shows a picture of one small portion of the bottom of #1 microcrater taken with a scanning electron microscope (Yamakoshi 1990).

In Figs. 8.7 and 8.8, X-ray microanalyses give us the K-X ray pictures from Fe and Al materials. The dark regions of Al are occupied by many Fe dots.

Fig. 8.7. Fe-X ray microanalysis of the bottom region of the microcrater.

Fig. 8.8. Al-X ray microanalysis of the bottom region of the microcrater.

It is difficult to estimate quantitatively the residual amounts of the projectile material which lined the inside wall of the microcrater. At first, soft and ductile materials such as Au or Al seemed very suitable target material to catch incoming particles.

Recently developed studies are described in detail in the section on the "Comet Sample Return Project".

Foil stack catcher

We carried out some interesting laboratory experiments for cometary dust collection. In Kyoto University, Fujiwara (1978) has constructed a well-designed two-stage light gas gun, whose highest speed can reach up to 6 km/sec (≤1 g of projectile). In the first experiment of this study, we prepared an Al-foil stack target: 20 sheets of 15 μm thick Al foil were folded and stapled at three points at a 120° opening angle for each sheet. Finally, 10 stacks were fixed 5 mm apart, and the final stopper was a Mg plate, whose thickness was 5 mm. The stacks and the stopper were fixed in a ring-frame using a set of fringe.

The projectiles were basalt grains weighing about 0.1 g and mounted on nylon sabots. The experiments were carried out in a chamber which was kept in a vacuum by a rotary pump. A series of experiments were performed in which the projectile speed was about 1 km/sec.

After the impact, the projectiles were largely broken by the Al-foil stacks, but the final Mg stopper plates were scarcely scratched. About 78% of the initial projectile mass was collected from the Al stack house.

The foils were broken so greatly that the speed of the projectile decreased. In these experiments, the shock-waves ran into two dimensions at right angles toward the impact direction on the respective foil-surfaces.

As an analogy, the phenomena closely resemble heavy ionized particles, in that when they go through materials, the particle goes more slowly, and the energy loss rate becomes larger.

The broken holes became bigger and bigger, and near the very end the broken holes were extremely large. However, the final stopper, the Mg plate, was scarcely damaged.

Another experiment was carried out for comparison. The target was an Al-plate, whose thickness was 3.0 mm, exactly the same thickness of aluminum used in the previous foil-stack experiment. The projectile speed and other conditions were kept the same as those of the foil-stack experiment.

However, the Al-plate was penetrated with no trouble. Also, a crater was created in the final Mg plate stopper, whose depth was just 6 mm. And the crater and the target-hole materials were ejected away in opposite directions.

From these experiments, it appears that the Al-foil dust catcher is one of the most effective mechanisms for slowing down projectiles. A stack of many thin sheets of Al-foil is a "shock-absorber", and therefore it could be useful for the wall structure of space stations and/or lunar bases against hyperveolocity incoming cosmic debris. The damage to the foil stacks is shown in Fig. 8.10.

Fig. 8.9. Cross-sections of the foil-stack before and after the collision.

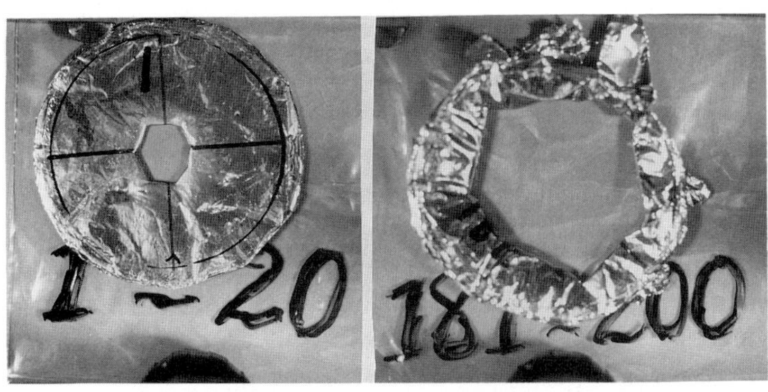

(a)

(b)

Fig. 8.10. (a) The surface and final stacks. (b) The perforated hole of the Al target of the same thickness and the Mg stopper (Yamakoshi 1990).

Dust sample capture devices on board a space shuttle

J. A. M. McDonnell *et al.* (1984) joined the third Space Shuttle "Pathfinder" OSS-1 Mission and mounted a passive catcher of double layer foil structure with 1 m^2 area. An upper layer of Al-foil 5 μm in thickness was held just 1.0 mm above a Kapton sheet by a Au-coated brass grid forming a capture-cell type of catcher.

The catcher was exposed in space by the Space Shuttle Mission for 8 days in March, 1982. The returned catcher was analyzed in the laboratories of the University of Kent, England.

McDonnell *et al.* (1984) showed topside and reverse pictures of an impact hole (about 20 μm in size) in the Al foil as well as the impact region of the Kapton sheet. The underside of the foil around the impact-hole was lined with sub-micron sized debris, which was scattered (splash-back) material from the shocked area on the Kapton.

Hypervelocity Impact Penetration in MFE Aluminium Foil.
(a) top side (b) underside

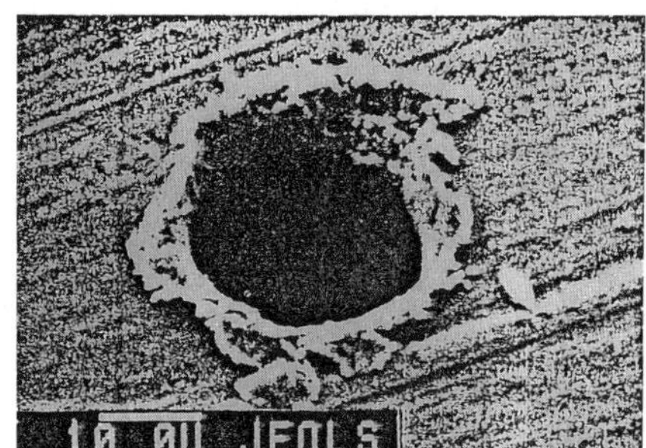

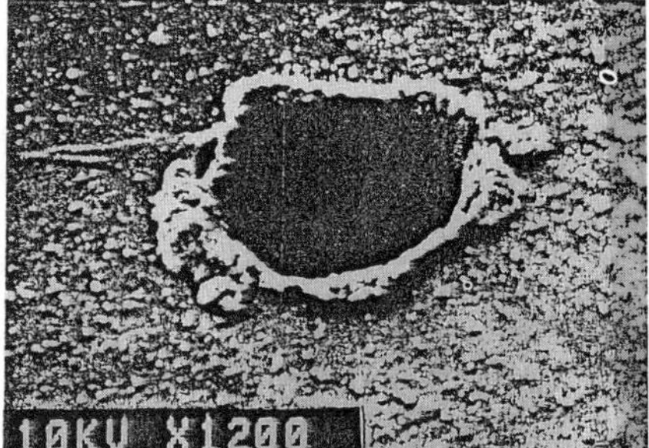

Impact Zone on MFE Kapton sheet.

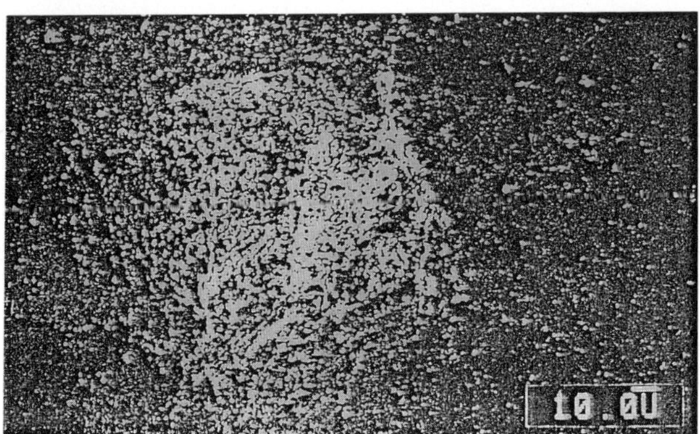

Fig. 8.11. Pictures around the impact holes (McDonnell *et al.* 1984).

They observed the underside of the Al foil sheets and found that the ejecta were roughly divided into two categories, globular and irregular. The sizes of the latter were slightly larger than those of the former. A sulfur component was detected in the globular grains, and an Fe component was found in the irregular grains.

McDonnell *et al.* (1984) concluded that this capture device is effective as a meteoroid damper.

REFERENCES

Fechtig H., Gruen E. T. and Kissel J. 1978, "Laboratory Simulation" in "Cosmic Dust" ed. J. A. M. McDonnell (John Wiley & Sons).

Hallgren D. S. and Hemmenway C. L. 1976, Proc. IAU Colloquium #31 Lecture Notes in Physics 48 284.

Humes D. H. 1976, "The Jovian meteoroid environment" "Jupiter" ed. T. Gehrels (Univ. Arizona Press) 1052.

Iglseder H. and Igenbergs E. 1987 Intern. J. Impact Engng. 5 381.

Igenbergs E., Huedepohl A., Uesugi K. T., Hayashi T., Svedhem H., Iglseder H., Koller G., Glassmachers A., Gruen E. T., Schwehm G., Mizutani H., Fujimura A., Yamamoto T., Araki H., Ishii Y., Yamakoshi K. and Nogami K. 1991 "The Munich Dust Counter" Proc. IAU Colloquium #126 "Origin and Evolution of Interplanetary Dust" eds. A. C. Lavesseur-Regourd and H. Hasegawa (Kluwer Acad. Publish. Tokyo) pp. 15.

McDonnell J. A. M., Carey W. C. 1984 Nature 309 237.

Simpson J. A. and Tuzzolino A. M. 1985 Nucl. Instr. Meth. A236 187.

Yamakoshi K., Fujiwara A. and Asada N. 1985 J. Geomagne. Geoelectr. 37 987.

Yamakoshi 1988 unpublished.

Yamakoshi K. 1990 "Impact-Collision Studies" Proc. Workshop on Space Debris ISAS Report SP 11 39.

Yanagisawa M., Sato K., Hara T., Furuya K., Uchida M., Yamori A. and Kawashima N. 1987 ISAS-Report #51 (in Japanese)

Chapter 9

Microcrater Formation by Dust Impacts

9.1 Dust accelerators

In order to simulate the dust impact collision phenomena in many laboratories, we need a "dust-accelerator", whose ability is required to create similar conditions of space in the laboratories, to accelerate an artificial grain up to the actual velocity.

In this section, three methods to accelerate projectiles up to hypervelocity in laboratory simulation experiments are reviewed: railguns, two-stage light gas guns and electro-magnetic accelerators for fine grains.

Railguns

Railguns are currently used on military satellites and are intended for aggressive purposes such as SDI (The Operation "Star Wars" of the USA). Railguns are designed to fire a projectile whose mass is more than a few grams, and velocity is aimed up to 10~20 km/sec.

As shown in Fig. 9.1, an electric current and a magnetic field at right angles to the current produce a repulsive force. The projectile is equipped with plasma-armatures at the bottom. These armatures slide inside the barrel, between two rails, keeping contact with the rails.

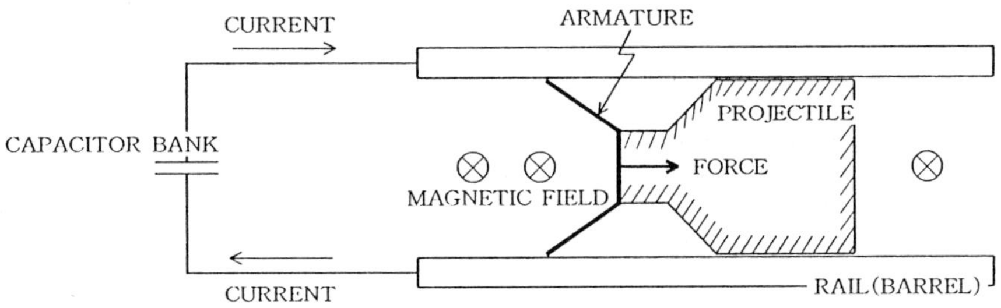

Fig. 9.1. A schematic cross-section of the railgun.

Since the acceleration limit of a "two-stage light gas gun" is forecast to be 5~6 km/sec, the railgun mechanisms have been developed from the end of the 1970's.

Railguns are applicable to large-scale collisional experiments, or even to military purposes in space. Yanagisawa *et al.* (1987) obtained the relation between the charging voltage of the capacitor tanks and the projectile velocities; preliminary results using their gun, located in a laboratory of the ISAS (National Institute for Space and Astronautical Sciences), are shown in Fig. 9.2. Improvement of the gun is now in progress.

In the future, it will be possible to "launch" spaceprobes or satellites themselves by large-scale railgun devices, instead of rocket promotion.

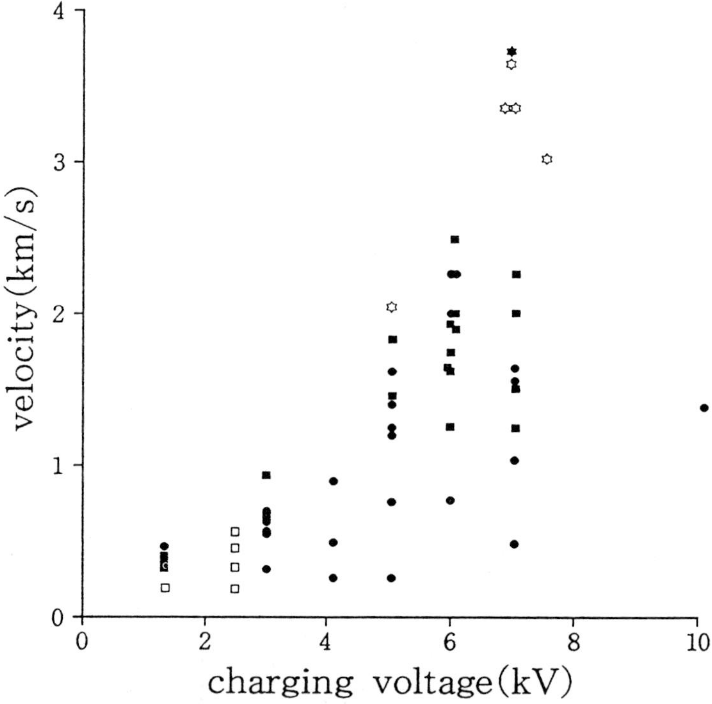

Fig. 9.2. A relationship between velocity and mass of a projectile (Yamori *et al.* 1987). The projectile velocities are plotted against the charging voltage of 6 mF capacitor banks. The open stars correspond to the round bore plasma-armature experiments. The solid star is for hybrid armature shot. The others represent the experiments with the square bore guns.

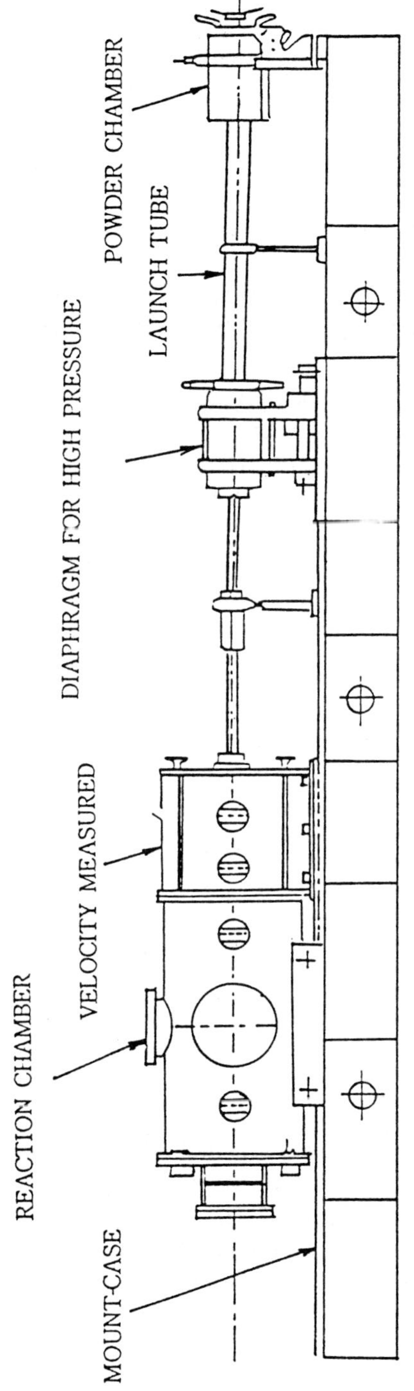

Fig. 9.3. A cross-section of the two stage light gas gun located in Kyoto University (Fujiwara 1978).

Two-stage light gas gun

Powder explosion in a powder chamber moves a polyethylene piston to press light gas, such as He gas, in a barrel tube. The He gas is pressurized so highly (and raised to such high temperatures) that a high-pressure-coupling diaphragm is split, and the gas jetting through the opened hole pushes the projectile, mounted on a sabot, up to 5~6 km/sec in final velocities.

The impact collision with a target occurs in a reaction chamber kept in a vacuum. With a slight time delay, the He gas jet follows the projectile and flows into the chamber.

At two positions between the diaphragm and the reaction chamber, the velocity of the projectile is measured by various methods, such as laser beam refraction using prisms. A cross-section of the two-stage light gas gun constructed by Fujiwara (1978) and located in Kyoto University is shown below.

Up to 15 g of smokeless powder is exploded, and the pressurized He gas is obtained as 5~6 kg/cm^2 in pressure. The mass of the projectile is less than 1 g. For example, a 0.21 g nylon sphere is fired with 7 g, 9 g and 10 g smokeless powder, so that final velocities of 2.5, 4.0 and 4.5 km/sec are obtained, respectively.

Sometimes the projectiles are enclosed in cylindrical plastic sabots during the launch cycle to prevent mutual abrasion of the launch tubes and projectiles. The sabots are split into two halves along their axes of symmetry, so that centrifugal forces induced by rifling in the tube cause the sabot pieces to deflect away from the projectile flight curve. A sabot carrying small particles is displayed in Fig. 9.4.

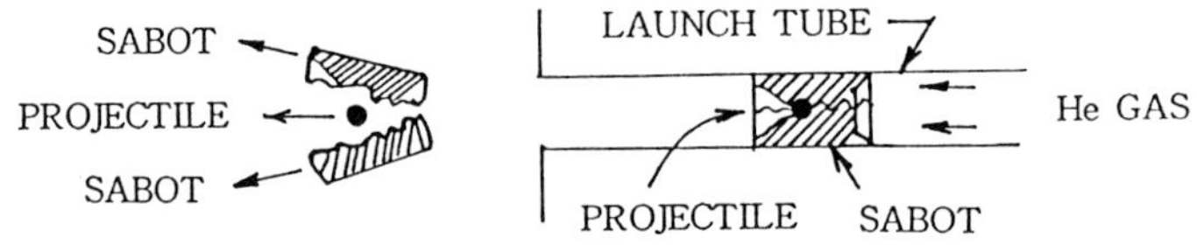

Fig. 9.4. Typical sabots for fine particles.

Electrostatic accelerators

In a Van de Graaff type reaction, kinetic energy is transferred to a projectile by electrostatic interaction. This mechanism applied to dust-grain acceleration was described by Friichtenicht (1962). In this acceleration, the projectile carries a positive charge and obtains kinetic energy as it goes through a potential difference, U. In Newtonian expression, the following equation is valid:

$1/2 \cdot mv^2 = q \cdot U$, where m = particle mass, v = particle velocity, q = particle charge, and U is the acceleration voltage.

It is evident that there are two requirements for producing high kinetic energies for a projectile: the charge to mass ratio, (q/m), of the projectile and the accelerator voltage should both be as high as possible, which is limited by the generator's ability. It is, however, difficult to put as much of the electron charge on the grain surface as possible.

In this acceleration mechanism, we have some fundamental relationships for a homogeneous sphere:

$m \propto r^3$, where r is the particle radius.

For the electric charge per unit area we have $q \propto r^2$.

Therefore, $r \cdot v^2 \propto U$, so that $r \cdot v^2 \propto mv^2/q$, then we have $r \propto v^{-2}$.

then
$$m \propto \mathrm{v}^{-6}, \quad q = \mathrm{v}^{-4}.$$

This means that the smallest particles will be the fastest as they have the higher charge to mass ratio. Fine metal grains, stored in a pocket, are released from the source pocket by applying certain electric pulses, and some of the grains are taken up at once to a projectile position.

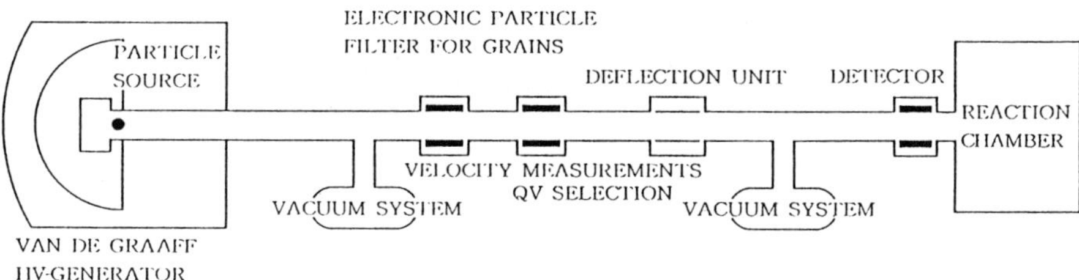

Fig. 9.5. A sketch of an electrostatic accelerator for metal grains.

9.2 [D/T] Parameter in Microcrater Studies

When we investigate features of cosmic meteoroids, such as size, mass, density and chemical composition, microcraters produced on the surfaces of spacecrafts and/or lunar rocks by hypervelocity impacts of cosmic micro-debris provide good clues. In addition, in the early period of space development by mankind in the 1970's it was expected that major parts of projectiles would be caught by target materials.

Brownlee *et al.* (1973) found two peaks on a diagram of (D/T) and the frequencies, where D is the diameter (inside rim) and T is the depth (inside rim) of the microcraters found on the surfaces of lunar samples. $(D/T) = 1.4$ and $= 1.9$ can be seen from Fig. 9.6.

Here a cross-section of a microcrater is shown; diameter *Di*, *Dr* and *T*.

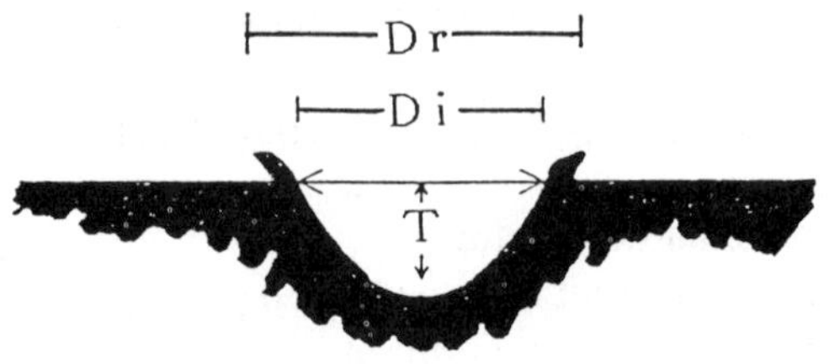

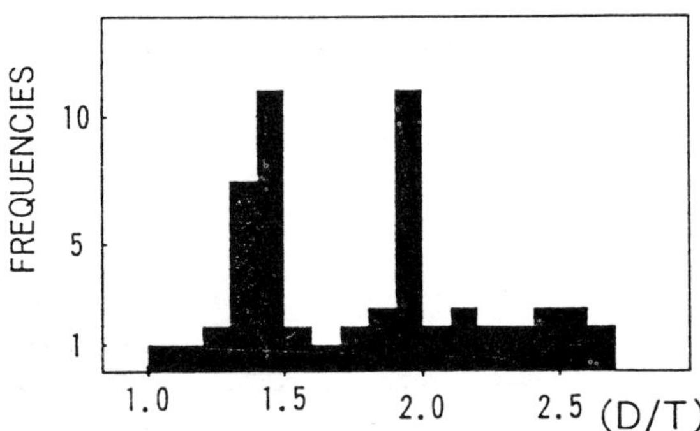

Fig. 9.6. The distribution of the (D/T) values of the microcraters found on the surfaces of lunar rocks (Brownlee *et al.* 1973).

Nagel and Fechtig (1980) performed simulation experiments in a laboratory and obtained the diagram shown below. These two peaks clearly coincide with those of the lunar results.

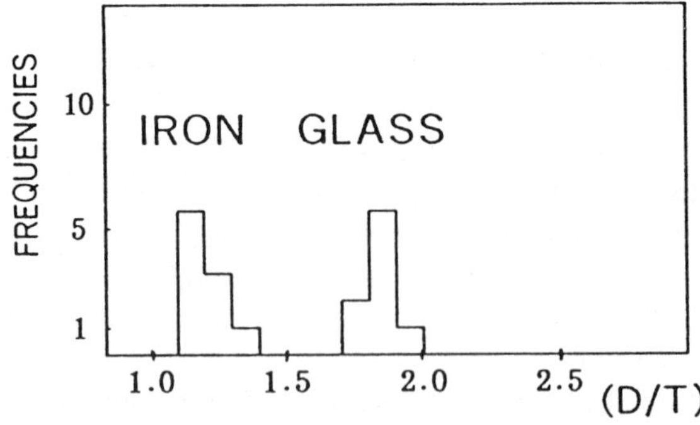

Fig. 9.7. (D/T) distributions obtained by simulation experiments using iron and glassy projectiles with feldspar targets (Nagel and Fechtig 1980).

The two peaks produced by iron and glassy projectiles in the simulation experiments are $(D/T) = 1.4$ for iron and $(D/T) = 1.9$ for glassy projectiles, respectively. These simulation results are obtained in the velocity range between 4 and 11 km/sec. We consider that these are remarkably important phenomena for researching meteoroid impacts.

Nagel and Fechtig (1980) showed that the (D/T) values are independent of the impact velocity of the projectiles. By these interesting results we have the possibility to evaluate the projectile materials, iron or stony rocks?

Detailed investigations of microcraters remaining on the surfaces of spacecraft after a long travel in space will show us the abundance ratio of (iron/silicate) dusts in space. This promises to be the most simple and non-sophisticated detector design for chemical composition analyses of the cosmic dust in space. Yamakoshi and Fujiwara (1986) added the results and confirmed the independence of the parameters (D/T) of the projectile velocities, which are shown below in Fig. 9.8.

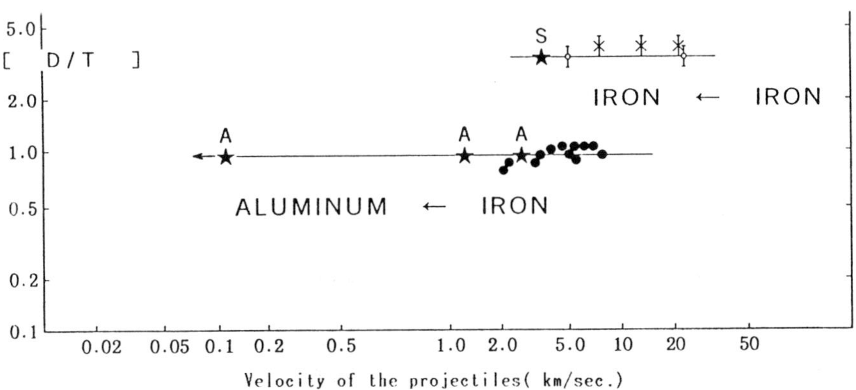

Fig. 9.8. The diagram of (D/T) values vs. the velocities of the projectiles are shown [MPI in Heidelberg; o = iron, × = carbon projectiles to stainless steel targets and • = iron projectiles to aluminum targets] and [ours; *S = stainless steel projectile to stainless steel target and *A = Fe (94%) + Ni (6%), projectiles to aluminum targets].

It is very interesting that the (D/T) is independent of the velocity in a wide range of projectiles and is dependent only on the combination of the materials of the projectile and target.

The fundamental physical process of microcrater formation and (D/T) ratio determination in various combinations of materials must be studied in the near future.

LDEF (long duration exposure facilities)

Particulate detection in the near Earth's orbit was performed with LDEF (long duration exposure facilities) rotated for 5.778 years. It was launched on 7th April 1984 and recovered on 12th January 1990. During the flight, it had remained about 458 km in mean altitude and rotated the Earth at 28.5° inclination with a velocity of 7.64 km/sec (McDonnell *et al.* 1991). LDEF has 130 m^2 of surface area and an open-grid cylindrical structure on which a series of rectangular trays used for hardware were attached. These trays faced in 14 directions (12 directions along the sides and 2 directions at the end) (Zolensky *et al.* 1991).

Many scientific projects are concentrated into this LDEF, such as "UHCRE" for ultra heavy cosmic ray experiments,. "SDIE" for space debris impact experiments, "STS" for impact erosion damage in space, "MAP" for microabrasion experiments, "IDE" for interplanetary dust experiments and so on.

Bernhard *et al.* (1991) found on the recovered LDEF's surfaces 174 individual microcraters from an Al plate (>99% pure, #1100 Al) and 180 craters from the sheet metal flanges (6061-T6 Al) of two instrumental trays. Over 50% of the craters have no measurable chemical residues.

Warren *et al.* (1991) counted 238 craters on a gold plate whose area was ~1 m^2. The sizes of the craters ranged between 10 and 400 μm. They used an empirical relation, $Dr = 1.3\ Di$, where Dr is a diameter referring to "rim-crest-rim-crest", D = diameter inside the rims.

Zook (1991) reported the crater distribution on the surface of LDEF in Table 9.1.

Table 9.1. Top-Bottom and Top-Side crater number ratios vs. crater diameter, mass and slope (Zook 1991).

Crater diameter (μm)	Mass (g)	Slope	Top/bottom	Top/side
500	7.8×10^{-7}	−0.90	105	1.36
100	8.1×10^{-3}	−0.48	45	1.40
2	1.2×10^{-13}	−0.41	39	1.40

REFERENCES

Bernhard R. P., Brownlee D. E., Laurance M. R., Davidson W. L. and Horrz F. 1991, Abstracts 22th LPSC pp. 91.

Brownlee D. E. 1973, Proc. Lunar Planet Sci. (Pergamon Press) pp. 3197.

Fujiwara A 1978, Nature 272 602.

McDonnell J. A. M., Sullivan K., Stevenson T. J. and Niblett D. H. 1991 A monograph of "Origin and Evolution of Interplanetary Dust" eds. A. C. Levasseur-Regourd and Hasegawa H., Kluwer Academic Publish. pp. 3.

Nagel K. and Fechtig H. 1980 Planet. Space Sci. 28 567.

Warren J., See T. H., Cardenas F., Laurance M., Messenger S., Brownlee D. E. and Hoerz F. 1991, Abstracts of 22th LPSC pp. 1465.

Yamakoshi K. and Fujiwara. A 1986 Annual Rep. ICRR, Univ. Tokyo ICRR-Houkoku-67-88-1 pp. 271 (in Japanese).

Yamakoshi K., 1990 "Impact Collision Studies" Proc. Workshop on Space Debris ISAS Report SP 11.

Yanagisawa M., Sato K., Hara T., Nakazawa T., Furuya K., Uchida M., Yamori A. and Kawashima N. 1987 ISAS- Report #51 (in Japanese).

Zook H. A. 1991, Abstracts of 22th LPSC pp. 157.

Chapter 10

Searches for Traces of Meteoroids Accreted in Soil and Deep-Sea Sediment Layers

10.1 Recovered Particles from Crater Regions of Meteorite Falls

Although magnetic (iron, stony) spherules were discovered by Murray and Renard (1891) during the deep-sea expedition by Challenger VI in 1872~76, their extraterrestrial origin was not proved until the 1970's. During the long time duration after their discovery, some efforts were made to prove their cosmic origins. In these examinations, meteorites and/or collected debris from the vicinity of meteorite craters (in almost all cases, the large-scale craters were made by iron-meteorite falls) were chosen as the reference standards.

In this section, some studies of the tiny debris which were gathered as solidified droplets derived from the dust-trails scattered from the ablated surface of falling meteorites are reviewed.

The most famous iron-meteorite crater in the world is the so-called "Canyon Diablo" crater in Arizona, USA, whose age is estimated as about 50,000 years.

The fallen and scattered materials were gathered on an industrial scale. The fragments broken up during the entry into the Earth's atmosphere or the recoiled fragments or secondary ejecta formed during the crater formation process have been studied. An extensive report was compiled by Vdovykin (1973) in which he described in detail the debris recovered from the crater region.

Mead *et al.* (1962) analyzed metallic spheroid samples collected from the Canyon Diablo crater region with an electron microprobe analyzer. They described some spheroids consisting of a fine-grained granular kamacite core, commonly intergrown and surrounded by a geothite, $Fe(OH)_2$, like iron oxide and maghemite. Associated with the kamacite exists an interstitial pink mineral inclusions, which is optically similar to the pink schreibersite, $(Fe, Ni)_3P$ or $(Fe, Ni)_2P$, in metallic spherules and tektites. Their result are given in Table 10.1.

Table 10.1. Chemical compositions of the spherules analyzed with XMA (Mead *et al.* 1962).

Sample	Mineral	No. of analyses	[Fe] (%)	[Ni] (%)	[P] (%)	[S] (%)	Total (%)
#1	Kamacite	8	98.3	2.3	—	—	100.6
	Light gray oxide	6	61.6	5.7	—	—	67.3
	Dark gray oxide	5	51.7	3.0	—	—	54.7
#2	Kamacite	9 for Fe 7 for Ni	74.1	23.5	—	—	97.6
	Dark gray oxide	2	33.3	5.1	—	—	38.4
#3	Schreibersite	8	41.3	40.7	19.0	—	101.10
	Dark, gray oxide	16 for Fe 12 for Ni	45.3	2.2	— —	— —	47.5
#4	Pink	10 for Fe 9 for S 6 for Ni	43.7	6.4	—	24.6	74.6
#5	Kamacite	3	82.5	12.8	—	—	95.3
	Dark gray oxide	4	42.1	8.6	—	—	50.7
#6	Light gray oxide	5	59.0	5.1	—	—	64.1
#7	Pink	10	53.4	16.0	—	28.6	98.0

Hodge *et al.* (1963) analyzed the magnetic spheroids obtained from volcanic erupted materials, Greenland ice layers and meteorite craters of Canyon Diablo and Shikhote Alin. The comparative data are shown in Table 10.2.

Table 10.2. The analyzed data obtained from volcanic material, ice layers and meteorite craters (Hodge *et al.* 1963).

(% by weight)

Sample	No of samples	[Al]	[Si]	[Ca]	[Ti]	[Mn]	[Fe]	[Ni]	[K]	[Mg]
Volcanic origins	50	7	15	3	7	0.2	31	0	1	0
Ice layer contained	43	2	9	2	0.5	0.2	41	0.5	0.01	1
Crater region	17	2	6	1	1	0.2	48	2	0.2	0

These data can be considered to be typical mean and also bulk compositions of the spheroid samples from volcanos, ice layers and iron meteorite craters.

Ehmann *et al.* (1968) analyzed spherules collected from the Ballinger (Canyon Diablo) crater and obtained [Au] = 3.4 ppm and [Ir] = 5.6 ppm.

In the "parent" iron meteorite, Canyon Diablo, the bulk composition of [Au] = 1.43~2.08 ppm and [Ir] = 1.8~2.5 ppm, so that some grades or types of enrichment of the elements possibly occurred during the formation process of the spherules. The difference of the chemical composition between the parent body and the ejecta gives us a hint to search for fundamental processes which occurred at the instant of the impact.

However, the possibility that the spherules analyzed by Ehmann *et al.* (1968) originated from other meteorites or cosmic dust itself cannot be neglected.

Vdovykin (1973) discussed the correlation between the chemical composition of the parent body of the iron meteorite "Canyon Diablo" and the magnetic spherules gathered from its crater. The mean compositions of the magnetic spherules (oxygen component was not determined) were: [Fe] =53.77 %, [Ni] = 9.06 %, [Co]=0.95 %, [P] = 0.56 %, [S] = 0.96 %, [CaO] = 1.10 % and [acid residues] = 16.10 % (in weight).

Metallic cores of the magnetic spherules gathered from this crater region were: [Fe] = 75.63 %, [Ni] = 17.30 %, [Co] = 2.22 % and [acid residues] = 1.5 %.

The parent body, Canyon Diablo, has the bulk chemical components: [Fe] – 92.3 %, [Ni] = 7.25 % and [Co] = 0.49 %. So we have [Ni/Fe] – 0.168 for an average value of the magnetic spherules, 0.228 for the average value of their nuclear cores and also (7.25/92.3) = 0.0796 for their bulk composition.

In the impactites, the nickel content sometimes exceeded 45 % or more; Nininger (1968) tried to explain these enrichments of siderophile components as happening during the process of evaporation of the iron meteorite by impact shock and recrystallization as well as melting of meteoritic materials.

However, his hypothesis was criticized by some scientists, because if the spherules were formed by the condensation of the vapoured gases due to collisional impact, the chemical compositions would be homogeneous.

Spherules must have relict minerals from their parent meteoroids; however, the changes of mineralogical compositions may be caused by high temperature and sudden cooling.

Dense (1971) did not study the changes of chemical composition of scattered meteoroids from a meteorite crater, however, rather studied the impactite formation around and inside the crater; their compositions were in general the mixed products of the soil or rock components of the impact site as well as the fractions of fallen meteorites. The low temperature fractions are selectively melted, and during condensation and cooling a grain absorbs the vapourized gas and/or melted fractions.

On February 12, 1947, a large iron meteorite fell in the Maritime, Siberia, as a meteoritic shower; it was named "Sikhote Alin". Krinov (1964) reported on the debris gathered in the vicinity of the craters made by "Sikhote Alin". The fragments are classified into two groups: the first group includes unbroken individuals formed when the original material disrupted. Each fragment has a fusion crust and regma-glyptic relief. The second one includes splinters of individual fragments of the first group. The "splinters" vary in appearance due to the fusion crust or degree of deformation, sometimes having sharp, curved edges. Surface soil samples weighing more than 1.5 kg were taken from the interior of the crater area. The magnetic fraction was gathered from the soil samples, and then magnetic particles were picked out.

Krinov (1964) classified the particles gathered in the "Sikhote Alin" crater region into three categories:

(1) particles of tiny, highly angular and irregular forms,

(2) a rarer component of spherules and spheroidal forms and

(3) very rare fragments from the aerial disruption of the original meteorite, which have a complete fusion crust.

He named these three categories in his paper as (1) meteoritic dust, (2) meteoric dust and (3) micrometeorites, respectively. However, these namings are used incorrectly, I think, because these scientific terms are recognized as having essential meanings which are different for each other, concerning their origins, sizes and phenomena.

The category (2) samples contain three types of particles (a) spherules, (b) liquid droplet forms and (c) flask forms.

The particles of liquid droplet forms have an oval form with a pointed tail. The particles of flask forms are hollow with spheroidal shapes and/or grooved necks on the outside. They are further divided into three groups: typical forms, cap forms and spherical forms, shown in Fig. 10.1.

Fig. 10.1. Various types of irregular forms (Krinov 1964): [liquid droplet form], [typical flask], [cap form flask], [spherical flask].

These liquid droplet and flask-type particles look just like the debris with the following history: Melted surface material forms droplet and the connecting necks between the body surface, and the created neck becomes narrower and narrower and finally separates from the surface. However, they cool so suddenly that the separation edges of the connecting necks are solidified and preserved (see Section 6.6).

Krinov (1964) reported on the size distribution of the spherules gathered from the vicinity of the crater. However, they could not prove that the all of them were derived from separate fractions from the same parent meteorite.

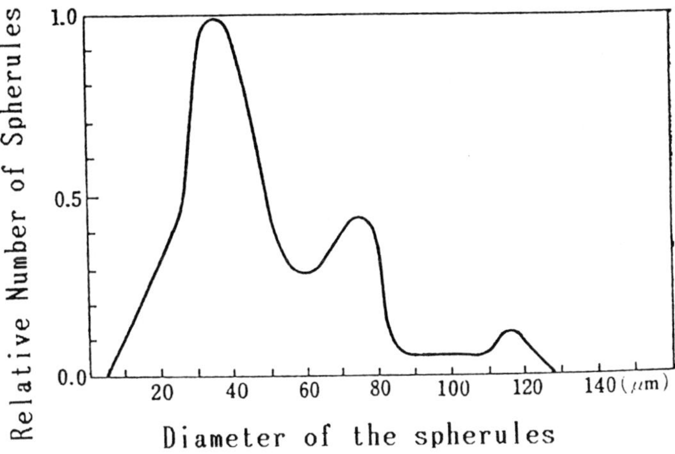

Fig. 10.2. Size distribution of spherules gathered from the Sikhote Alin crater region (Krinov 1964).

In Fig. 10.2 it is not clear whether the three maxima correspond to distinct materials and/or mineral phases, or are simply caused by poor statistics.

The Tunguska Event happened on June 30, 1908, with an enormous explosion over an isolated forest region in Siberia. A few decades after the impact time, research scientists could find only large scale destruction of the forest in this region but no traces of cosmic meteoroid fall. They could discover neither broken meteoroids nor craters.

In the expedition performed by L. A. Kulik and his coworkers starting in 1921, they discovered craters due to meteoritic falls. In 1958 an expedition team was reorganized by the Committee on Meteorites, Academy of Science, USSR. In this region a large volume of soil samples was collected and sieved, and some types of spherules were gathered (Kirova 1964).

In the description by Kirova (1964), the spherules gathered from the soil samples were divided into two groups: magnetic spheroids, and silicate spherules and other fused forms. As the transitional forms, researchers identified droplets, retorts, shell-like forms and others. The spherules were sometimes porous and earthy. And other forms, such as thin-walled shells, thin-walled earthy spherules with sievelike structures, droplets, bright forms with noses and tails, pear shaped and egg-shaped, broken subspherical forms and irregular ones were found.

The silicate spherules ranged from transparent to opaque and sometimes looked like oval-shaped grains. The size distribution of the magnetic and silicate spherules are shown in Fig. 10.3.

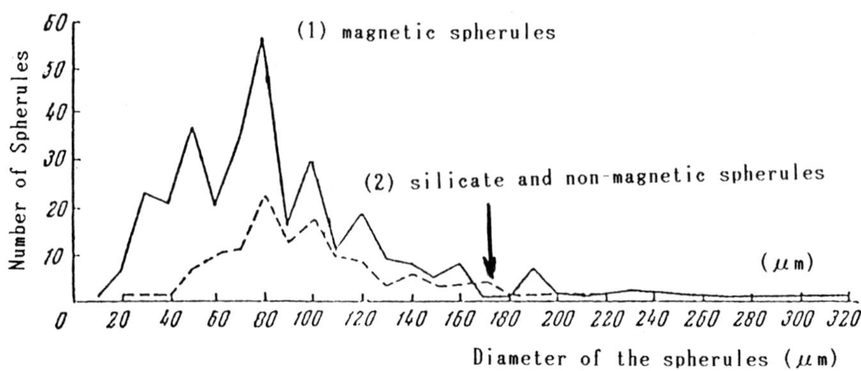

Fig. 10.3. The size distribution of the collected magnetic and silicate spherules; (1) magnetic, (2) silicate and non-magnetic spherules (Kirova 1964).

Ganapathy (1980) reported the chemical composition of magnetic spherules discovered in the "mysterious" region of Tunguska, Siberia. He analyzed several spherules with INAA. The results obtained are shown below:

Table 10.3. The chemical composition of eight magnetic spherules gathered from the Tunguska region (Ganagathy 1980).

Sample	Weight (μg)	[Ni] (%)	[Co] (ppm)	[Ir] (ppb)	[Au] (ppb)	[Cr] (ppm)	[Sb] (ppb)
A*	43	0.062	153	25	11 ± 4	76	1900
C	69	0.170	676	69	8 ± 15	5 ± 2	—
D	60	0.063	107	55	≤5	8 ± 2	210
E	74	0.036	115	29	17 ± 5	9 ± 1	200
G	36	0.093	233	45	7 ± 6	11	640
F**	59	6.07	2580	2500	216	3770	≤130
B#	111	4.63	4280	9040	—	393	—
H##	55	21.4	11900	56900	420	—	≤290

* contained 4.4 ppm As ** 53 ± 55 ppb Re and 3.8 ± 1.2 ppm Os # 430 ± 280 ppb Re, 34 ± 4 Os and 40 ppm Pt ## 4300 ppb Re, 85 ± 5 ppm Os and 80 ppm Pt.

(The experimental errors given by Ganapathy are so small, that most have been removed by the author for simplification).

He pointed out that the Au content correlates with the nickel content fairly well; however, he never analyzed spherules not related to the Tunguska meteorite, because his Ni-Ir diagram showed a good correlation. In general, cosmic meteoroids, which fall down daily and everywhere, are ablated micrometeorites and do not show any such good correlation between Ni and Ir contents, because the Ir content varies independently of the Ni content in cosmic meteoroids so much. He reported that in the Antarctic ice layer of 1908 A.D. an Ir-enriched one was found, and he stated that it will be a good age-mark for geological studies.

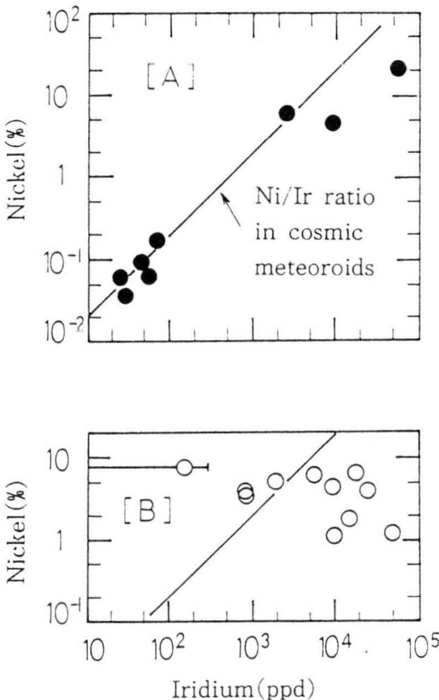

Fig. 10.4. The Ni-Ir correlation diagrams of magnetic spherules [A] shows Tunguska samples, [B] shows Ni/Ir ratios in ordinary spherules accreted daily onto the Earth's surface (Ganapathy 1980).

10.2 Meteoroids and Impactites Reveal Hidden Astroblems

This is an inverse of the story mentioned in previous sections. Meteoroids and impactites can be collected from the vicinity of fall-craters, and the gathered debris can be analyzed and studied from various points of view. Bird's-eye view pictures taken from airplanes sometimes reveal hidden craters, in which forests have already grown or water has accumulated and changed to fresh water lakes. Sometimes they look just like dead volcanic calderas.

However, impactites and/or cosites contain high contents of siderophile elements. Searches for hidden astroblems (meteorite carters) have been an attractive subject for the last two decades.

In southern Germany, the Ries crater region is considered to be an astroblem. Morgan *et al.* (1979) researched the collected spherules and analyzed their chemical composition. In these cases, almost all the iron grains were already oxidized and could barely be collected using magnetic collectors. Morgan and his coworkers analyzed noble metal elements, Ir and Os, and their correlation curve is drawn below. They stated that the gradient of the curve represents the cosmic abundance, and it is much different from the ratio of terrestrial samples.

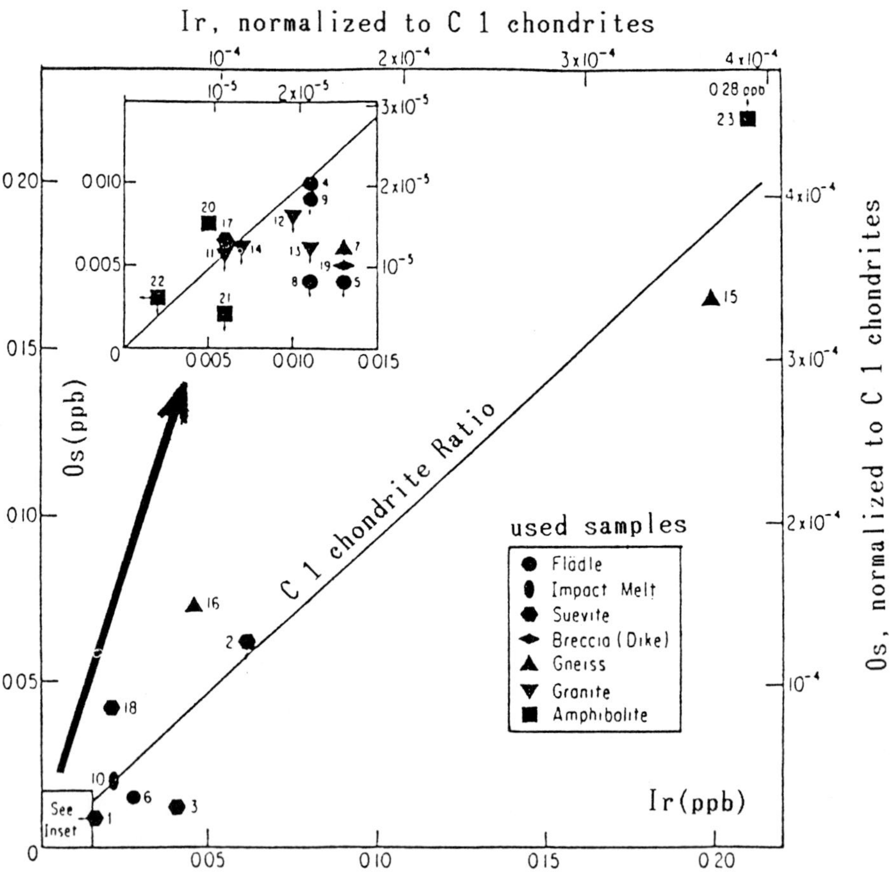

Fig. 10.5. Noble metal contents of the samples collected from the vicinity of Ries crater region (Morgan *et al.* 1979).

Table 10.4. Major element composition of basement rocks from Laeppajaervi.

Element\Samples	L3/1	L2	L5	L7	L3/2-2	Accuracy
SiO_2 (%)	68.10	—	73.9	50.4	62.70	±2%
Al_2O_3 (%)	15.0	15.3	14.5	17.0	17.6	±3%
MgO (%)	2.50	1.2	0.4	5.8	3.12	±3%
FeO (%)	4.10	3.46	0.16	11.1	6.7	±2%
MnO (ppm)	504	547	49	1673	678	±2%
TiO_2 (%)	0.59	0.46	—	0.60	0.70	±5%
CaO (%)	2.11	4.16	0.50	9.2	0.69	±2%
Na_2O (%)	3.13	3.87	2.29	3.50	1.46	±2%
K_2O (%)	2.16	1.42	7.84	5.0	4.80	±5%
Total (%)	98.14	98.6	99.6	103.1	97.84	

The Laeppajaervi crater in Finland has been studied using the Rb-Sr dating method. Elemental compositions of impact melt and basement rocks were determined by Reimold (1982).

The Rb-Sr dated age shows 77 million years, which corresponds to the Precambrian age. He analyzed the major and trace elemental compositions of basements of the region as well as the impact melt breccias. The results are shown in Tables 10.4~10.7.

Table 10.5. Trace element abundances of basement rocks from Laeppajaervi (Reimold 1982).

(ppm) Element\samples	Mica schist L 3/1	Mica schist L 3/2-2	Grano diorite L 4	Granite-pegmatite L 5	Mphibolite L 7	Accuracy (±%)
Sc	11.0	20.3	11.6	0.38	34.3	5
V	75.8	137.0	39.9	—	310.1	
Rb	130.9	216.9	65.4	161.4	14.9	15
Sr	250.6	118.4	281.9	130.5	158.8	
Zr	222.1	157.3	155.5	21.8	—	
Mo	—	3.47	—	—	0.9	
Cs	7.5	12.0	1.34	1.91	0.49	20
Ba	363.5	841.0	351.7	452.7	37.1	15
La	35.1	31.9	29.3	1.12	5.28	5
Ce	77.8	78.5	53.6	2.15	12.7	15
Nd	33.5	32.7	19.8	—	17.3	20
Sm	5.7	5.3	2.8	0.3	2.5	10
Eu	1.4	0.9	0.82	0.6	1.0	5
Tb	0.6	0.86	0.3	0.11	0.52	10
Dy	3.7	4.4	1.4	—	3.34	10
Ho	0.3	0.55	0.34	0.25	0.68	10
Tm	—	0.75	—	—	—	
Yb	2.77	2.4	0.82	0.54	2.46	5
Lu	0.41	0.32	0.13	0.07	0.35	10
Hf	4.6	3.5	2.64	0.3	1.5	10
Ta	0.9	1.0	0.94	0.01	0.23	15
Th	10.0	9.3	5.3	0.3	0.4	10
U	3.4	3.1	1.35	0.7	0.27	10
Cr	92.1	144.3	23.9	6.6	59.8	5
Co	10.8	12.5	8.7	0.3	36.9	5
Ni	61.0	63.9	16.6	9.1	52.0	15
Zn	86.7	122.7	53.6	4.0	126.9	10
Ga	18.1	22.6	19.7	7.7	18.4	10
As	29.0	104.5	0.3	0.95	0.51	15
Se	0.004	—	—	—	—	
Br	—	0.42	1.14	1.1	—	
W	—	5.1	—	—	—	10
Ir(ppb)	<0.5	<1.0	—	<0.5	<0.5	20
Au(ppb)	1.4	2.6	<0.5	<0.5	2.75	15
Sb	—	0.12	—	—	0.29	

The major, minor and trace elements of the impact melt rocks or impactites which were gathered at the place of Laeppajaervi were measured using mass-spectrometry and RNAA (Reimold 1982).

Table 10.6. Major element compositions of imact melt rocks from Laeppajaervi. (Reimold 1982)

(% by weight)

Element	[La42]	[L1/9]	[L12/2]	[L14/4]	[L15/2]	[L1/14]	[L2/2]	[L1/9]	[L1/6]	(Accuracy)
SiO_2	66.8	67.0	67.7	68.2	65.4	67.7	66.10	68.6	66.3	±2%
Al_2O_3	17.9	18.2	15.3	14.3	16.0	15.3	18.5	16.6	15.8	±3%
MgO	2.4	2.9	1.1	2.9	3.0	2.1	1.75	1.3	2.67	±3%
FeO	4.9	5.0	6.1	5.5	5.1	5.0	5.29	3.78	3.20	±2%
MnO (ppm)	563	503	531	587	588	484	506	145	275	±2%
TiO_2	0.59	0.60	0.55	0.54	0.49	0.51	0.62	1.03	0.59	±5%
CaO	2.27	2.83	2.84	2.59	2.76	2.43	2.39	2.71	1.83	±2%
Na_2O	2.88	2.82	2.87	2.86	2.87	2.82	2.88	2.84	2.18	±2%
K_2O	3.75	3.69	3.52	3.49	3.73	3.79	3.81	3.30	4.25	±5%

Table 10.7. Trace element abundances in impactites from Laeppajaervi.

(% by weight)

Element	[La42]	[L9/1]	[L12/2]	[L14/2]	[L15/2]	[L1/14]	[L2/2]	[L1/9]	[L1/6]	(Accuracy)
Sc	12.6	12.7	15.3	14.6	13.8	12.9	10.2	13.2	10.8	±5%
V	126.4	119.2	111.8	112.7	119.7	103.3	91.6	124.5	99.9	
Rb	100.9	96.9	107.1	115.3	97.4	100.7	99.5	125.5	131.2	±15%
Sr	223.7	231.3	245.5	223.7	227.3	229.7	242.9	261.8	239.9	±15%
Zr	211.5	177.7	299.9	151.8	169.0	195.4	211.1	—	185.9	
Cs	1.57	1.55	1.75	2.1	1.71	1.75	1.8	—	1.3	±20%
Ba	1151.9	1124.5	1034.7	931.2	1198.1	1123.2	1159.6	1065.8	1241.6	±15%
La	46.1	45.8	112.9	44.3	48.0	45.2	43.8	79.6	50.8	±5%
Ce	85.1	90.5	212.2	95.0	93.1	92.0	90.7	151.8	94.5	±15%
Nd	24.6	38.1	88.3	35.4	41.5	38.1	38.5	61.6	5.1	±20%
Sm	61.6	6.0	11.8	6.2	6.1	6.1	5.8	12.0	6.2	±10%
Eu	1.16	1.30	1.30	1.25	1.21	1.1	1.1	1.75	1.4	±5%
Tb	0.59	0.63	0.88	0.63	0.61	0.55	0.7	1.3	0.6	±10%
Dy	2.64	2.87	4.6	2.96	3.8	2.54	3.9	5.2	3.3	±10%
Ho	—	0.63	—	—	0.44	0.37	1.24	1.9	0.7	±10%

Tm	—	0.14	—	—	—	1.4	0.8	—	0.6	
Yb	2.11	2.09	1.50	2.1	2.0	1.9	1.9	1.7	1.6	±5%
Lu	0.28	0.29	0.26	0.28	0.28	0.26	0.24	0.32	0.24	±10%
Hf	4.04	4.6	5.6	4.5	4.8	4.6	4.5	7.2	4.9	±10%
Ta	0.7	0.76	0.89	0.72	0.72	0.8	0.8	—	0.8	±15%
Th	10.7	10.4	24.8	10.3	10.1	11.3	10.6	17.9	12.3	±10%
U	2.25	2.7	3.5	2.5	3.3	2.75	2.8	4.9	2.5	±10%
Cr	124.1	117.7	161.9	177.3	144.6	122.3	129.0	121.2	108.5	±5%
Co	19.7	18.5	27.3	29.2	23.9	16.7	22.7	7.8	7.4	±5%
Ni	196.7	195.2	339.2	313.2	329.0	227.1	265.1	n.d.	101.0	±15%
Zn	106.6	95.8	91.0	55.2	116.7	96.0	96.2	n.d.	111.9	±15%
Ga	19.5	16.6	21.0	16.2	16.6	16.2	13.1	11.7	13.2	±10%
As	0.64	0.47	1.4	0.39	0.56	0.47	0.54	—	0.59	±15%
Se	0.53	<0.4	0.25	0.68	0.20	0.77	0.58	—	—	
Br	—	2.13	2.3	3.0	3.0	4.0	4.2	3.7	—	
W	—	1.7	—	0.9	2.2	1.3	1.3	—	0.7	±10%
Ir (ppb)	7.6	6.5	8.5	11.1	9.6	5.9	12.0	—	<1.7	±20%
Au (ppb)	2.2	3.3	6.2	6.0	4.7	6.7	8.6	4.3	9.7	±15%

It suggests meteoritic projectiles collided to make a lake there and that the siderophile elemental contents in impact breccias containing 195~340 ppm of Ni and 6~12 ppb of Ir are much higher than those in rock mixtures containing 54.8 ppm of Ni and 0.5 ppb of Ir. And the impact melt breccias are formed under an ultra high pressure estimated as larger than 60 GPa.

Impact melt breccias and impactites studies give us fruitful clues to discover hidden, ancient craters. And studies on shock metamorphosed rocks should be "partly" developed by military nuclear explosion tests underground.

10.3 Cosmic Matter Accretion Events Found in Dated, Deep-Sea Sediment Cores

Iridium content anomalies at the K-T boundary have been studied by many workers. In the deep-sea sediment cores as well, enrichments of the siderophile components have been surveyed. Kyte *et al.* (1981) reported that RNAA and INAA methods were applied to Ir and Au content determinations of long sediment cores obtained from the Antarctic Ocean. A 230 Myr old layer contained remarkable Ir and Au contents; however, no mass extinction was reported there.

They found two large (1.5~2.0 m/m in size) grains in that sample. The two grains had 40% of Ir of the total Ir in the sediment sample.

Yamakoshi *et al.* (1991) examined two core samples collected by piston-corers on board the R/V Hakuho-Maru, Ocean Research Institute, University of Tokyo, and analyzed siderophile elements using INAA.

The sites of the collected cores were located at longitude 160°W, and sample A (KH-68-4, St-15) was obtained at 12°00'N (depth 5775 m) and sample B (KH-68-4, St-18) at 1°59' (5360 m). These two cores were fortunately dated by the paleomagnetic method (Kobayashi *et al.* 1971) as well as by the ^{10}Be method. Iridium contents and the ratios of (Co/Fe) in the dated, respective layers of the cores are shown in Fig. 10.7. The Ir values are expressed in the unit time interval (10^5 yr).

Each core sample was analyzed by INAA in a non-destructive form, since the detection sensitivity of Ir was not so good; however, if higher Ir contents were found, the gamma peaks due to the neutron-activated Ir would be distinguished from the background level of the detection limit (for sample A; Ir~1 ppb, sample B~0.5 ppb).

The neutron irradiations were carried out by a TRIGA II reactor, whose neutron flux was 0.7×10^{11} (neutrons/sec, cm^2) at irradiation times of 10 hours for sample A and 18 hours for sample B. The processed Canyon Diablo and JB-1 (Japanese Standard of Basalt #1) were used as the references for elemental composition determination.

A few layers of the sediment samples in the cores were lost, so that their Ir contents could not be determined.

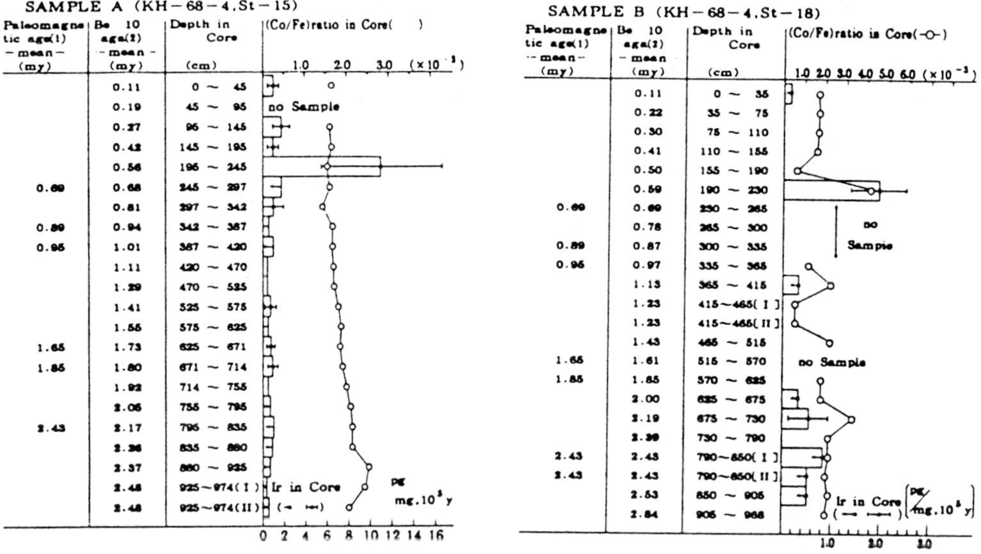

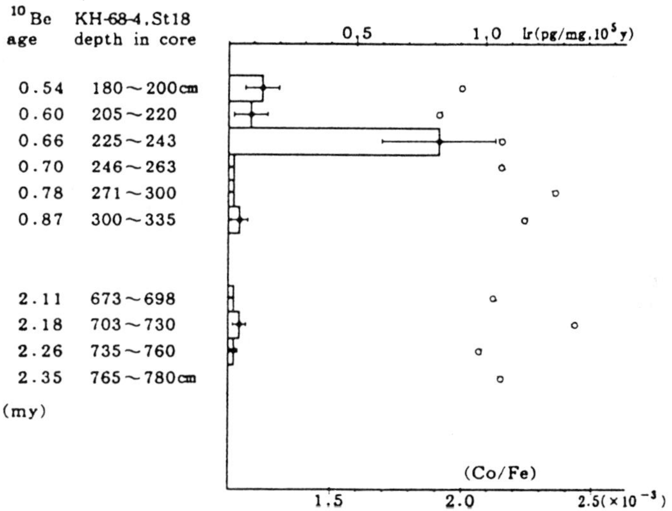

Fig. 10.8 Iridium contents and (Co/Fe) ratios in the dated cores obtained in the Central Pacific Ocean (Yamakoshi *et al.* 1991).

In the figures, the statistical errors of all (Co/Fe) points were much smaller than ±0.5%, so the error bars are not illustrated here. When the Ir content of $[-a \pm b]$ was obtained after calculation, if $[b - a] \leq 0$, the content was shown as 0. The statistical errors for Ir contents were taken as $\pm\sigma$.

In the figures, positive Ir anomalies at [0.58 ± 0.06] or [0.59 ± 0.05] My before present (BP) were found. It is considered that the peaks of Ir contents in two cores could be caused by the same event, which might have occurred at around 0.58 My B.P. in the central region of the Pacific Ocean. These anomalies might have occurred as a local event, so that no mass extinction was recorded in the history of the Earth (Yamakoshi *et al.* 1991).

Whether cosmic matter accretions have occurred at some time intervals in the past or interrelations between cosmic matter accretion and the dynamic revolution of the solar system could be discussed in the future.

REFERENCES

Dense M. R., 1971, J. Geophys. Res. 76 5552.
Grieve R. A. F. and Robertson P. B., 1979, Icarus 38 212.
Ehmann W. D. Baedecker P. A. and McKown D. M., 1970, 34 493.
Hodge P. W. and Wright F. W., 1964, Nature 225 717.
Ganapathy R., 1980, Science 209 921.
Kirova O. A., 1984 Ann. New York Acad. Sci. 119 235.
Krinov E. L., 1964, Ann. New York Acad. Sci. 119 224.
Kyte F. T., Zhou Z. and Wasson J. T., 1980 Nature 288 651
Mead C. W., 1965 American Geologist 50 667.
Morgan J. W., Janssens M. J. Hertogen J., Gros J. and Takahashi H., 1979, Geochim. Cosmochim. Acta 43 803.
Murray J. and Renard A. F., 1891, Deep Sea Deposite Report Challenger VI Expedition Chap. 4.
Nininger H. H., 1956, Arizona's Meteorite Crater 88.
Palme H., Rammensee W. and Reimold U., 1980, 11th Lunar Planet. Sci. Conf. 848.
Reimold W. U., 1982, Geochim. Cosmochim. Acta 46 1203.
Yamakoshi K., Nogami K., Omori R., Misawa K. and Ma Jianguo, 1991 Proc. IAU Colloq. #126 "Origin and Evolution of Interplanetary Dust" eds A. C. Levasseur-Regourd, Kluwer Acad. Publish, pp. 53.
Vdobykin G. P., 1973, Space Sci. Rev. 14 758.

Chapter 11

Concluding Remarks

In Chaps. 1 to 10, the author described many experiments and problems concerned with cosmic dust in interplanetary space. Since cosmic dust is usually defined as a meteoroid, whose size or mass is smaller than those of meteorites, the origins and sources of the dust have not been elucidated that well. Recently, the dust whose mass range is between 10^{-16} and 10^{-6} g can be analyzed and captured by space-borne devices. The characteristics of cosmic dust, such as trajectories, velocity, mass and major compositions can be determined by space-borne analyzers. Dust catchers to gather cometary dust ejected from periodic comets are being planned and developed.

In addition, we can obtain valuable information by these projects about extrasolar, primordial dust, beta-meteoroids and so on. The dust whose size range is between 0.1 and 100 μm is gathered by impact-plate collectors on an airplane in the stratosphere. (The weight of a spherical dust whose size is 1 μm and density is 1 g/cc can be estimated to be 0.52×10^{-12} g.)

Sometimes we call them "Brownlee's particles". They are curated by NASA Johnson Space Flight Center. These dust samples are thought not to have suffered so much thermal degeneration during their passage of the Earth's atmosphere because of their small sizes. Fraundorf *et al.* (1982) calculated the increased temperatures of the dust which enters into the atmosphere.

Hunter and Parkin (1960) calculated the critical sizes of the penetrating meteoroids in the Earth's atmosphere at which they cannot be melted as 15 μm for iron and 75 μm for stony ones. In general, if a dust particle radiates as a black body and loses no mass from molten material during the heating in the Earth's atmosphere, its highest temperature is given roughly by the formula:

$$\rho \mathrm{v}^3 = 8\sigma T^4,$$

where ρ is the air density, v is the velocity of penetrating dust, and σ is the Stefan-Boltzmann constant.

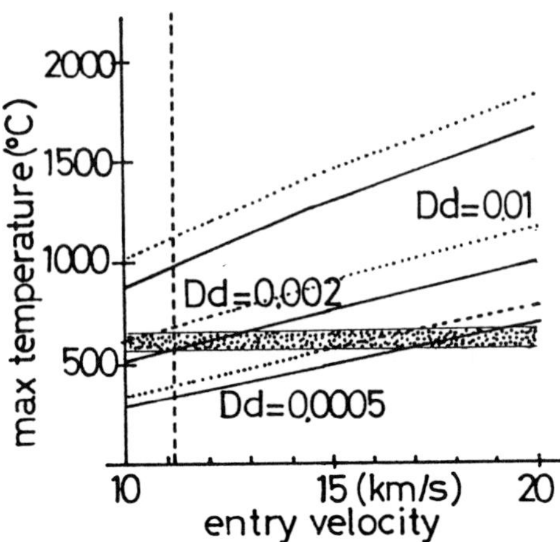

Fig. 11.1. The increased temperatures of accreting dust with the angle of 0° (normal; dashed curve) and 45° (solid curve) (*D* = size in cm, *d* = density). The dotted region shows the temperature at which the nuclear tracks are annealed in chondritic meteoroids (≤600°C). A vertical dashed line represents the escape velocity from the Earth.

The dust larger than a few hundred μm is changed to "cosmic spherules", which can be found and collected from deep-sea sediments and polar ice layers. The pre-atmospheric forms of the spherules will be irregular just like so-called "micro-meteorites".

Because of the meteor-flashing, the initial volume of the micro-meteorites may be decreased to 1/2~1/20, so that refractory elements are enriched and volatile ones are evaporated away due to the thermal characteristics of the elements in the stony or metallic parent bodies.

Love and Brownlee (1991) studied over 50 000 cases of the micro-meteoroid grains whose sizes ranged between 10 μm and 1 mm. The incoming velocities were assumed to range between 11.2 and 72 km/sec.

These grains attained the highest temperature at 85~90 km in altitude, and the duration time is estimated as ~1 sec. The submillimeter sized grains rarely exceeded 1700°C. They stated clearly that half of the grains which survived after meteor-flashings, whose sizes are larger than 70 μm, decreased to 1/3~1/2 of their pre-atmospheric mass. These results agree well with spherule-data obtained experimentally: remarkably abundant native iron, gold and REE contents in spherules. And they are consistent with the results of higher values of ($^3He/^4He$) ratios in magnetic fractions smaller than 74 μm gathered from deep-sea sediments. (See Chapter 6 in the text.)

In studies of "cosmic spherules" several problems now remain unsolved:

(1) the formation process of platinum nuclear cores and nuggets inside the spherules is unclear,

(2) fruitful comparison of spherule size distributions with the estimated results of meteor observations can be expected,

(3) possibility of finding primordial and/or extrasolar materials that we have,

(4) precise determination of cosmic ray induced nuclides which are carried by cosmic matter accreted onto the Earth's surface must be performed,

(5) chemical interaction and fractionation of the spherules with deep-sea water (weathering) and so on,

(6) simple comparisons in chemical composition between spherules and meteorites may give us special information on "meteoritic spherules" only, and we have a dangerous possibility to miss "terrestrial (fractionated) spherules".

The smallest sized dust fractions can also be studied with deep-sea sediments. Noble gas investigations revealed unmelted dust grains implanted with solar flare particles which were gathered into the smallest size group with the strongest magnetic property of deep-sea sediments. In the near future, various information will be obtained from these studies.

Circumstellar (extrasolar), interstellar (Galactic) and intergalactic (extra-Galactic) dusts will be studied by optical and infrared observations, and the obtained data will be compared with the laboratory simulation results.

The interrelation between interstellar and interplanetary dusts is one of the most important subjects now. And also the study of the sources (supplying origins) and sinks (captured by planets, sublimation and evaporation) of interplanetary space should be an eternal subject for us. The best clue to investigating extrasolar materials is to find isotopic anomalies in various elements due to nuclear syntheses different from those given for the solar system.

In this article, the author emphasized intentionally the studies of interplanetary "meteoroid samples". Astronomical observations and optical emission or absorption analyses of heated dust clouds in galaxies, experiments and theories of grain formations have been not described so much here; wish thus to write about such subjects in the future.

REFERENCES

Fraundorf P., Flynn G. J., Shirak J. and Walker R. M. 1980 LPSC XI 297.

Hunter W and Parkin D. W. 1960, Proc. Roy. Soc. A 255 382.

Love S. G. and Brownlee D. E. 1991, Icarus 89 26

Appendix

During the editorial and publication processes of this book much newly-obtained information on extraterrestrial dust came forward. Some of these developments are described here now.

A. Direct Measurement of Dust-Swarms and Dust-Trails by Dust-Detectors on Board Space-Crafts in Interplanetary Space

Results obtained from observation by the dust counter on board the Japanese space-craft "HITEN", launched January 24, 1990, came forth. The observation period was extended to March, 1993. Some of the results are contrary to what we had expected:

(1) no observation was made of impact-ejecta from the lunar surface, as had been expected. These should be detectable around the time a meteorite falls onto the lunar surface, just as has been detected with lunar meteorites found in Antarctica. However, this fact might be attributable to the counter's short period of observation of the lunar surface;

(2) no observation was made of increases in the dust flux at the Lagrange points, which are the equilibrium points of gravitational forces between the Earth and the Moon. The point moves day by day, dependent on the relative positions of the Earth and the Moon. At these points we had expected to find dust swarms of low velocity;

(3) we had expected the dust counter to measure dust swarms and streams originating from cometary trails. Unfortunately, the dust counter on board of HITEN space-craft did not record any remarkable increases in the dust count, which would have indicated their presence;

(4) we observed a dust grain whose velocity exceeded 70 km/sec. It was considered to be of extra solar origin.

From October 27, 1990 (1.03 AU) to August 24, 1991 (4 AU from the Sun) the dust detector on board the spacecraft Ulysses registered 50 large impacts over this period of 301 days (~0.5 impacts/day). Initially, the impact rate was about ~0.15 impacts/day, which is compatible with data obtained by Galileo's detector at the same distance from the Sun. Galileo's detector registered 377 large impacts during 489 days, from December 28, 1989 (0.88 AU) to May 1, 1991

(0.7–1.45 AU) (Gruen *et al.* 1992). The detectors on board the Ulysses and the Galileo have about the same dimensions and characteristics. The dust flux measured by Ulysses' detector is dependent on the heliocentric distance, whereas, within the Asteroidal belt (<2 AU) the flux is independent of this distance. Within the Asteroidal belt the average flux increased slightly, but remarkable variations (up to down) were observed.

Also, Ulysses' detector encountered a dust stream at 5.4 AU and at a distance from Jupiter 540 times the planet's radius. The background rate recorded at this position was ~0.4 impacts/day, but on one day (March 10, 1992) 126 dust impacts were registered. This remarkable, locally concentrated, dust trail is apparently a dust stream that Ulysses had encountered, which might have derived from either Asteroidal objects or cometary trails. The IR satellite, IRAS, has observed dust trails concentrated near the orbits of short period comets in the past (Sykes *et al.* 1986), and many researchers have predicted such trails for P/comets also to be present (Keller *et al.* 1991).

Gruen *et al.* (1993) reported that the spacecraft "Ulysses" changed her orbit and inclined at ca. 80° to the ecliptic plane of Jupiter on February 8, 1992. Ulysses then encountered dust-bursts with duration times from 4.7 to 43 hours in (28 ± 3) day periodic intervals. The dust streams appeared to have originated from the direction of Jupiter. During this period the number of the impacts ranged from 3 to 124 particles, their mass range was 0.1–90 ($\times 10^{-15}$ gms), their velocity range was 20–56 (km/sec), the impact sites from Jupiter were 553–1980 R(J). Gruen *et al.* (1993) made the following hypotheses with respect to the origin of the dust streams:

(1) the narrow and collimated streams must have originated from a near source;

(2) the paths of the dust streams are collimated on a position near Jupiter!;

(3) all streams came from the direction to Jupiter;

(4) the observed periodic intervals suggest a single dust source and exclude the possibility of accidental coincidence of cometary dust trails or asteroidal dust clouds.

They also discussed the acceleration of the particles and the change in angle of their motion due to the interaction of the charged dust particles with the interplanetary magnetic field.

B. LDEF Sample Analyses

The satellite LDEF (Long Duration Exposure Facility) was retrieved in January 1991 by the space-shuttle "Columbia" after having drifted around the Earth 5.7 years at an altitude of 400–500 km. The body surface of LDEF, which has 12-sided cylindrical form, 4.3 meters in diameter, 9.1 meters in length and a surface area of ~130 m^2 and which is constructed from aluminum, stainless steel, gold and aluminized teflon sheet (a thermal blanket), had been exposed to cosmic dust, artificial debris, cosmic radiation and atomic oxygen in the upper ionosphere (see Section 9.2).

Two aluminum clamp samples taken from the LDEF, parts "A07, C03" and "A08, C03", were given to us by NASA in August 1992. On these samples 16 microcraters were found, whose sizes were 130 × 80 mm and 8 mm in thickness each.

In this work on the LDEF samples, the authors have employed geometrical and chemical analyses on the microcraters for the purposes of determining projectile's composition (Yamakoshi *et al.* 1993). Long-lived, cosmic ray produced radioisotopes will be measured using a low background radiation counter for determination of the average altitudes of the satellite LDEF both during its mission and during the time it was drifting around the Earth.

It is well known that the ratios of the diameter (inside the rim, D) to the depth (from surface level, T), (D/T), is dependent only on the combination of targets and projectile's materials and independent of the projectile's velocity, size and weight (Nagel and Fechtig 1980; Yamakoshi 1990). From simulation experiments on aluminum targets using iron and plastic projectiles, it was found that (D/T) ≒ 1 and ≒ 2, respectively. From Fig. A-1 it may be seen that (D/T) ≒ 1 for iron projectiles to aluminum targets and (D/T) ≒ 3.5 for iron projectiles to iron targets according to results obtained in simulation experiments and that these measures are valid for wide ranges of the projectile's velocity; however, the data for aluminum-iron combinations ranged between 0.92 ~ 1.0. Nagel and Fechtig (1980) determined that, when aluminum targets are used, $(D/T) = C \cdot \rho^{-0.43}$, with ρ being the projectile's density and C being a constant. In our simulation experiments (Yamakoshi 1990) we obtained (D/T) = ~1.0 for aluminum targets with iron (ρ = 7.8) projectiles. So using the experimental relation mentioned above we have obtained C = 2.42, which is a normalization constant.

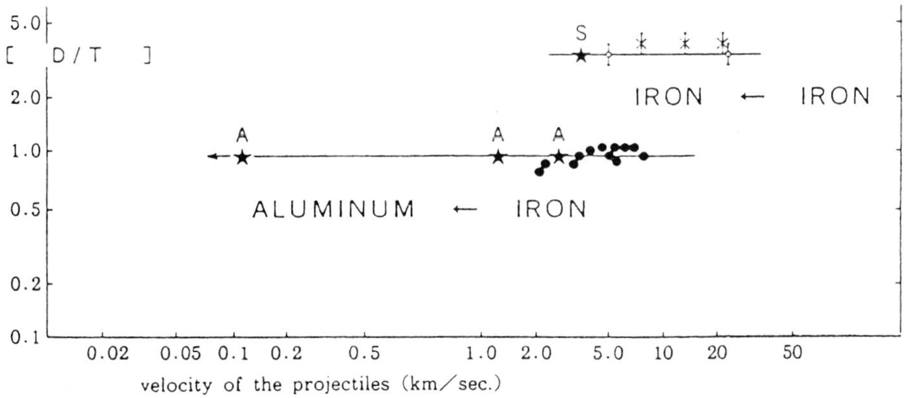

Fig. A-1. Diagram of (D/T) values vs. projectile's velocity (Yamakoshi 1990).

Using this formula, several cases of the projectiles are calculated and shown in Table A-1.

Table A-1. The calculated (D/T) values obtained with aluminum targets and various projectile's densities using the formula of (D/T) = 2.42 · $\rho^{-0.43}$ (Nagel and Fechtig 1980).

Projectile's density	(D/T) value	Remarks
7.8	1.0	iron
5.2	1.19	magnetic, iron spherules
3.8	1.36	chondrites (H)
3.4	1.43	carbonaceous chondrites
3.12	1.48	chondritic spherules (Murrel *et al.* 1980)
2.7	1.58	aluminum, granites, calcites
1.0	2.42	ice or fluffy grains
0.90	2.53	polystylene, nylon

Table A-2. Dimensions of the microcraters of LDEF samples.

Sample code		Diameter D; (μm)	Depth T; (μm)	(D/T) ratios	Remarks
[A07, C03]	A	457.16	203.27	2.14	stony
	B	51.035	25.80	1.98	stony
	C	74.383	12.84	5.79	residue? or small angle projection?
	D	112.020	51.31	2.18	stony
	E	123.45	111.37	1.11	iron
	F	34.028	12.90	2.64	plastics
	G	313.13	92.99	3.37	residue?
	H	282.27	71.12	3.97	residue?
[A08, C03]	J	118.76	56.76	2.09	stony
	K	138.661	33.27	4.17	residue? or small angle projection?
	L	185.53	96.60	1.91	stony
	M	152.463	71.56	2.13	stony
	N	339.81	118.53	1.80	stony
	O	139.624	57.53	2.43	icy
	P	369.71	114.54	3.23	residue? or small angle projection?

Nagel and Fechtig (1980) also used iron grains of size 0.9 μm and 1.6×10^3 μm as projectiles against aluminum targets and obtained the same values of 1.0 ± 0.2 and 1.0 ± 0.1 for (D/T) ratios, respectively. Therefore, (D/T) ratios are not influenced by the size and weight of a projectile.

In this work, the dimensions of 16 microcraters found on LDEF aluminum plates were measured using a laser-microscope (Model 1LM21, courtesy by Kobori 1993). The data are provided in Table A-2. A ratio of (D/T) $\geqq$ 2.4 is

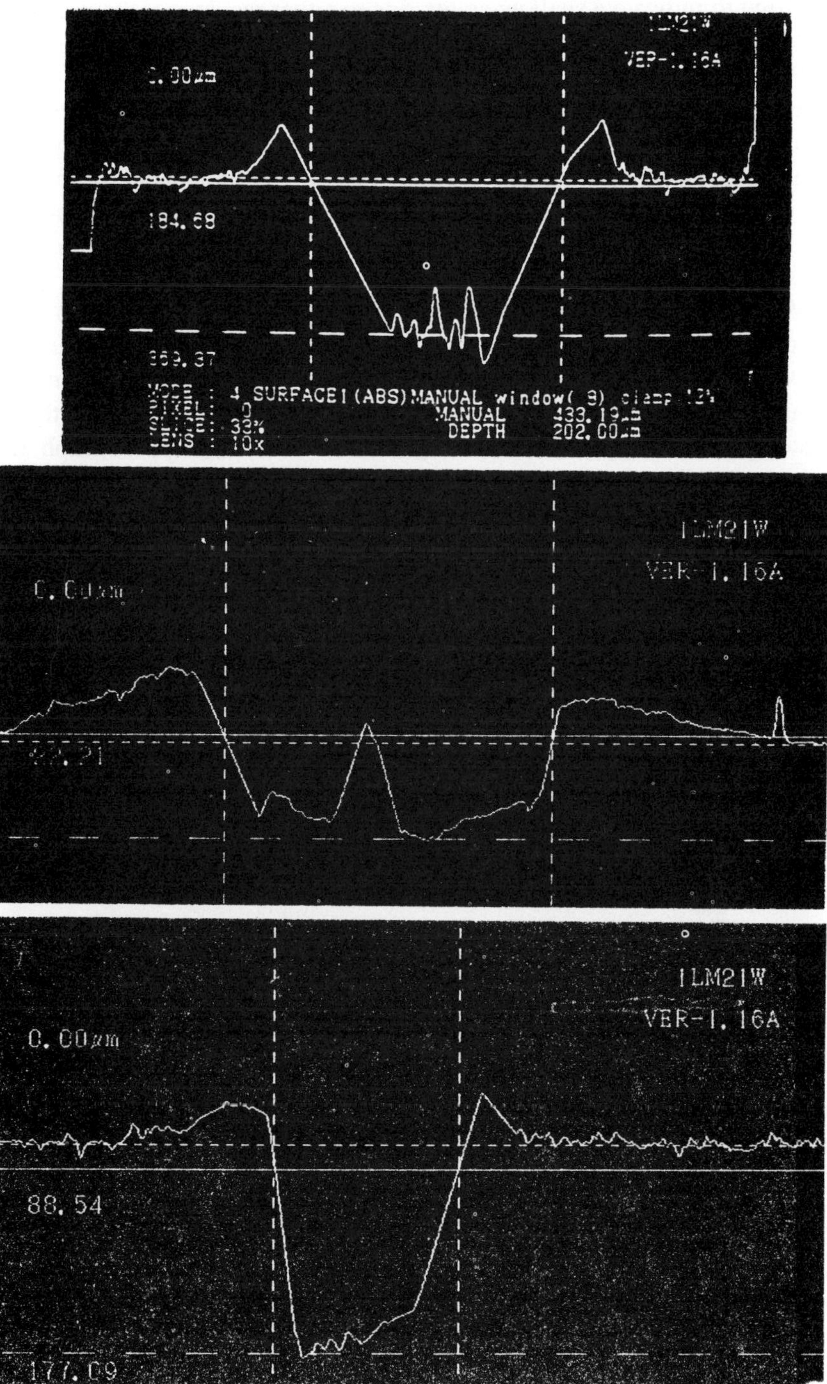

Figs. A-2[A], A-3[C] and A-4[E]. One dimensional cross-sections of the craters.

considered to indicate a shallow crater, due to either residues of the projectiles at the bottoms and/or the small angle of projection of the grain on impact.

Projectile residues appear, in the cross-section images taken by the laser-microscope, as irregularly shaped structures at the bottom of the shallow craters. These are shown in Figs. A-2, A-3 and A-4. The cross-section images are so irregular that further investigation of the residues in the craters has been done using an X-ray microanalyzer, the JEOL, Superprobe-JXA 8600M. One typical result of such an analysis is given in Fig. A-4. Here the geometrical data suggested that the projectile was an iron one.

We could deduce the projectile's materials from the (D/T) data of microcraters found on the LDEF samples. For one of the craters, we found (D/T) ≒ 1, which would suggest it had been caused by a cosmic iron grain impact. This is confirmed by X-ray investigation of the materials lining the inner surfaces of the microcraters. In Fig. A-5 a bird's-eye photograph of the microcrater (#N) is presented.

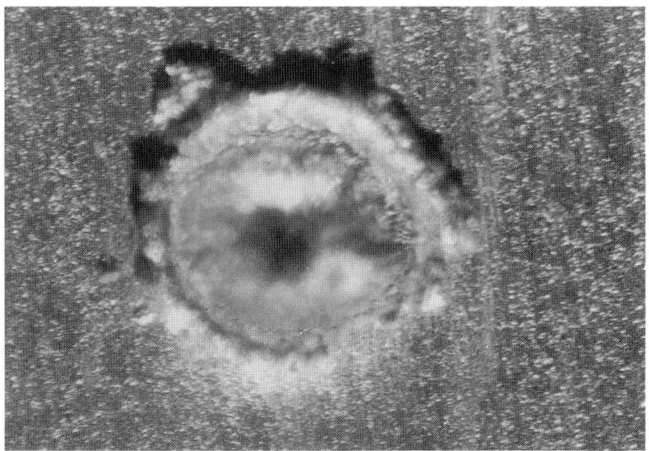

Fig. A-5. A bird's-eye photograph of the crater #N.

The (D/T) method has proven to be effective in determining a projectile's density as being roughly that of iron(7.8), chondritic(3.5) or ice(1.0). It is worthy to be studied, (D/T) values depend on whether density or hardness of the projectiles. However, it can not determine a projectile's origin, whether it be cosmic or artificial. Further investigations into this matter are now in progress.

In a newsletter of the LDEF mission, dated on August 15, 1992, a surprizing grain formation process was reported coming out of the Thermal Control Surfaces Experiments (TCSE). These grains became liquid drop forms on Teflon materials, occurring within an area 3 mm in width and 5 mm in length, whose area was exposed to active oxygen and solar ultra-violet radiation. The grain may be only several microns in size and they can be easily removed from the base-materials. The growth process is not yet understood, however, it is possible that the mechanism of grain formation in space, which has been unknown to us up to the present time, will be discovered.

C. Impactites and Noble Metal Particles Found at Geological Boundaries

Already Kyte found porous and rounded particles at a boundary having a high Ir content. These particles are now called "Kyte particles" (Kyte *et al.* 1981). In the 7th Symposium on Antarctic Meteorites (August, 1992; Tokyo) many interesting results were presented, among them the discovery of peculiar grains, which are considered to bear high content of Ir, the K-T boundary (6.5×10^7 y before present) at Morgan Creek, Canada (Ping Kong *et al.* 1992). One of them has a composition of 90% Rh, 5% Os and 5% Ir. The enrichment process of such noble metals into grain (~5 microns in size) has not yet been clarified.

Micro-tektites (65–220 microns in size) were found in loess at the Brunhes-Matsuyama boundary (Li Chun-lai *et al.* 1992), whose age was estimated to be 0.72 Ma. The authors of this project divided them into three groups:

(1) normal microtektites, whose chemical components include SiO_2, Al_2O_3, FeO and MgO, but which are largely variable in their exact composition. They are very similar to those present in deep sea sediments;

(2) bottle-green microtektites, which are largely composed of SiO_2 and have a higher MgO component. They are similar to those found in the Ivory Coast and Australia;

(3) high aluminum microtektites, which contain primarily SiO_2 and Al_2O_3. These are similar to those found in the deep sea cores at Molino de Cobo, Spain.

Unfortunately the authors did not discuss the most interesting subject, the correlation between the reversal in the geomagnetic field and impactite (micro-tektite) concentration at the same boundary. Is it possible that a large scaled meteoroid impact triggered the reversal of the geomagnetic field?

D. Magnesium Isotopic Composition of Deep Sea Spherules

Recently mass spectrometry of Mg isotopes (^{24}Mg = 78.60%, ^{25}Mg = 10.11% and ^{26}Mg = 11.29%) in chondritic, deep sea spherules was carried out (Misawa *et al.* 1992). Only one significant mass fractionations of up to 2 per mil, per amu was found (Sample Code; HWI#5). This one showed a small but notable non-linear ^{26}Mg isotopic effect ($\delta^{26}Mg$ = 1.20 per mil). The results are presented in Fig. A-6. Mass fractionation effects of Mg for the spherules are obtained within ±2 per mil per amu instrumental resolution bands. This small excess was also found in a study on a Brownlee's particle (Esat *et al.* 1979; see Section 4.3). In a recent laboratory simulation study (Misawa *et al.* 1990) it was shown that Mg mass fractionation of 2 to 30 per mil per amu is undoubtedly produced when a spherule mass loss of 20 to 85% occurred due either to mass loss from parent meteoroids by vaporization or to ablation effects occurring during entry into Earth's atmosphere.

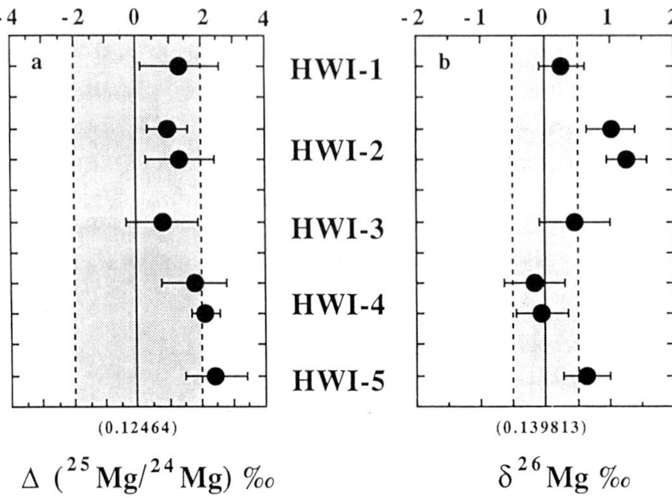

Fig. A-6. Mg isotopic components of five chondritic spherules from deep sea sediments: (a) mass fractionation effects of Mg for the spherules are within 3 per mil per amu. $\Delta(^{25}Mg/^{24}Mg) = \{(^{25}Mg/^{24}Mg)m/(^{25}Mg/^{24}Mg)s - 1\} \times 1000$, where m is the measured raw ($^{25}Mg/^{24}Mg$) ratios during the set, s is the grand mean value. (b) $\delta^{26}Mg = \{(^{25}Mg/^{24}Mg)c/0.139813 - 1\} \times 1000$, where c is the value for all corrected for fractionation.

In addition to the Mg problem in chondritic spherules, the cosmogenic ^{26}Al content is also important for understanding their origins. The decay products (^{26}Mg) of the primordial ^{26}Al were not observed in the experiments mentioned above. However, the decay products of cosmogenic ^{26}Al are known to accumulate gradually in the spherules. A lot of the chondritic spherules have been picked up and a bulk sample was measured in non-destructive forms with an extremely low background gamma ray facility located at an underground station of Mt. Nokogiriyama.

The preliminary data obtained were as follows (Yamakoshi 1992);

$$22.5 \pm 16.7 \text{ (}^{26}\text{Al dpm/gm of spherules).}$$

The data were obtained from 41.64 mgm of spherules and the counting time was "only" 32850 min. The data were estimated as undersaturation in space. As mentioned in Section 7.1, the Al content in stony meteoroids increases through the ablation process of atmospheric entry.

A project is now in progress in which a large number of chondritic spherules will be gathered and ^{26}Al measurements will be carried out by separation of various size groups.

E. REE and Lithophile Elemental Compositions of Chondritic Spherules

Stony, chondritic spherules, found in deep sea sediments, are thought to be primitive solar system materials. It is believed that they underwent re-melting and partial volatilization during atmospheric entry, and have undergone alteration by sea-water. In order to evaluate the chemical features of the primitive components of these spherules, it is necessary to clarify the effect of atmospheric entry and sea-water alteration on their chemical compositions. Thus deep sea, chondritic spherule have been analyzed for REE (rare earth elements) by a direct-loading mass spectrometric isotope dilution method. Also many major, minor and trace elements have been determined by INAA. One spherule (#1F), which did not show deep sea alteration, possessed a flat REE ($1.8 \times$ CI chondrites) pattern with no specific anomaly (see Section 6.4). Relatively high Ni, Ir and Au abundances in several spherules confirm that they are of extraterrestrial origin. On the other hand, the altered spherules show highly fractionated, light REE enriched patterns with large negative Ce and minor negative Eu anomalies.

It is well known that light/heavy REE fractionations can be produced by gas/solid and solid/liquid partitioning of REE during nebular and planetary processes. In the former case, the abundance pattern of REE is a function of the elemental volatility. However, in the latter case, the abundance is a function of the REE atomic number or the REE ionic radius. Since the REE abundances for the spherules 1A, 1D and 1E monotonically decrease with increasing atomic number, nebular processes can not be responsible for the light REE enrichment. If these REE fractionations are indigenous and of pre-terrestrial origin, parent meteoroids may be of differentiated ones. The effect of alteration during residence on the sea-floor is the second possible cause of light REE enrichment of the spherules. The cerium negative anomaly in REE patterns in the marine environmental circumstance is generally considered to have been caused by oxidation of Ce^{+3} to the Ce^{+4} state in sea water and the incorporation of Fe- and Mn-oxides.

It is possible that most of the chondritic spherules studied in the work (Misawa *et al.* 1993) derived from unfractionated chondritic materials which were affected by deep sea alteration during their residence on the deep sea floor. In Fig. A-7 CI chondrite normalized lithophile element abundances for the deep sea chondritic spherules are shown. Figure A-8 shows CI chondrite normalized elemental abundances of the spherule #1F. A comparison is also made between CM and CV chondrites. Although moderately volatile lithophile elements are systematically depleted as compared with chondritic abundances, refractory lithophile elements become uniformly enriched, and their abundances are in good agreement with those of CM chondrites. The element depletion relative to CM chondrites is considered to be due mainly to volatilization loss (K, Rb and Au),

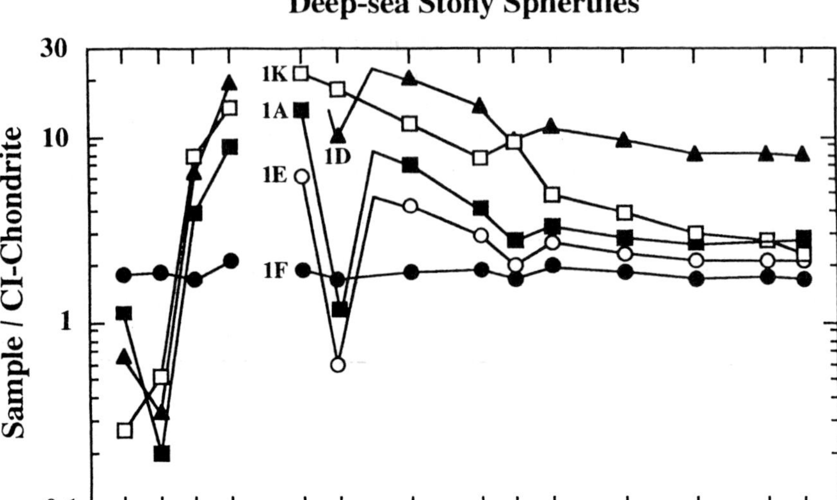

Fig. A-7. Lithophile element abundances for several deep sea, chondritic spherules, which are normalized with those of CI chondrites (Misawa *et al.* 1993).

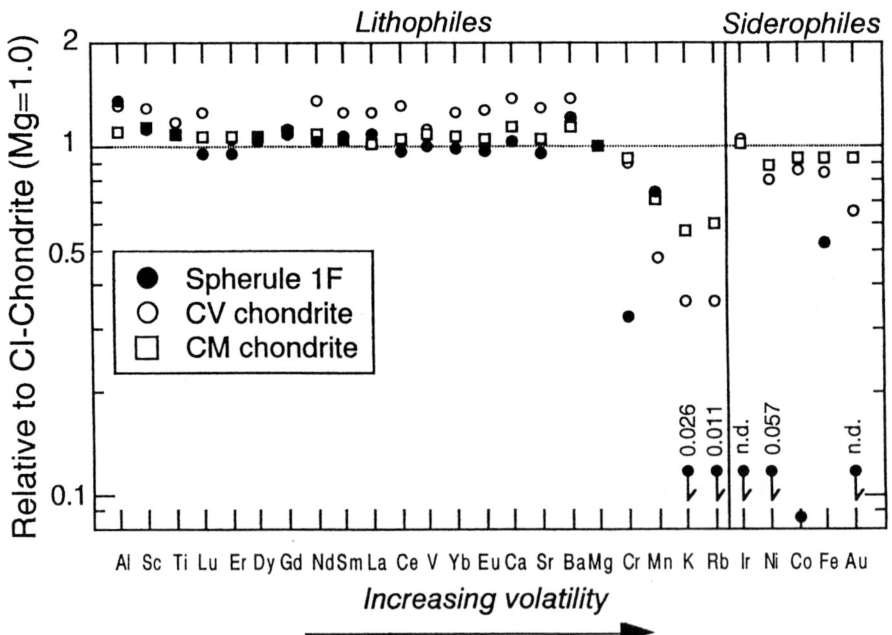

Fig. A-8. Lithophile and siderophile elemental abundances for the spherule #1F, which are normalized with those of CI chondrites. The spherule has a flat REE pattern with no specific anomaly. The elemental abundances are plotted in order of increasing volatility from left to right.

separation of siderophile components with Cr from the silicate melt, and the ejection of a metal or oxide droplet during atmospheric entry. In Table A-3 the results of chemical analyses by INAA and DL-MSID for several chondritic spherules are presented (Misawa *et al.* 1993).

Table A-3. Results of chemical analyses by INAA and DL-MSID of several deep-sea spherules.

Sample	1A	1D	1E	1F	1K
(μgm)	78.1	79.8	226.1	206.3	489.6
INAA Results					
Al (%)	1.2	1.4	2.1	2.1	4.6
Sc (ppm)	3.7	9.1	21.2	11.5	13.0
Ti (%)	< 0.4	0.28	< 0.3	nd	0.47
V (ppm)	814	545	98.6	99.6	124
Cr (%)	4.0	0.34	0.63	0.15	0.020
Mn (%)	0.19	0.10	0.20	0.26	0.078
Fe (%)	26.5	38.3	40.0	17.1	23.5
Co (ppm)	1100	504	1380	76.7	128
Ni (%)	0.86	nd	0.59	nd	nd
Ir (ppb)	2360	4470	nd	nd	nd
Au (ppb)	537	< 13	3890	nd	nd
DL-MSID Results					
Mg (%)	10.7	6.26	na	16.8	2.56
Ca (%)	0.182	0.300	na	1.64	0.465
Sr (ppm)	30.2	50.7	na	13.1	60.8
Ba (ppm)	21.0	45.5	na	5.04	33.5
La (ppm)	3.54	na	1.54	0.484	5.61
Ce (ppm)	0.763	6.52	0.375	1.10	11.7
Nd (ppm)	3.35	9.90	2.01	0.870	5.56
Sm (ppm)	0.626	2.28	0.444	0.290	1.17
Eu (ppm)	0.163	0.568	0.119	0.100	0.547
Gd (ppm)	0.674	2.32	0.542	0.404	0.983
Dy (ppm)	0.704	0.245	0.579	0.46	0.970
Er (ppm)	0.444	1.35	0.354	0.279	0.500
Yb (ppm)	na	1.35	0.353	0.290	0.466
Lu (ppm)	0.0713	0.207	0.0539	0.0427	0.0589

nd and na represent "not detected" and "not analyzed", respectively.

F. Interrelation between Extrasolar and Interplanetary Meteoroids

The isotopic anomalies of various elements occurring in primordial and abnormal samples, taken from primitive meteorites or interplanetary dusts, have been studied. Most of the isotopic heterogeneities have, up to now, been attributed to the incomplete mixing of the primordial and extra-solar materials and/or the mixing of materials of diverse nuclearsynthetic origins, such as spikes. Acid residues from primitive, unequilibrated meteorites, such as graphites, diamonds, silicon carbides and the refractory grains studied here (Kano *et al.* 1993a) are considered to be precursor materials which have survived the pre-solar nebula stages, and in these samples, the isotopic anomalies of noble gases, carbon, nitrogen, silicone, etc., have been found (Zinner *et al.* 1987, 1991a, 1991b; Kumar *et al.* 1991). The refractory, acid resistant residues from various meteorites [Allende (CV3), Nuevo Mercurio (H5), Canyon Diablo (IA)] are made by dissolving through a repeated and alternating treatment with HCl, HF and aqua regia at room temperature. They look just like composed aggregates. In Table A-4 the chemical compositions of the refractory acid resistant grains taken from various meteorites are shown.

Table A-4. Chemical Compositions of the Acid Residues of Allende (CV3), Nuevo Mercurio (H5) and Canyon Diablo (IA). The errors are the statistical errors ($\pm\sigma$) and nd denotes "not detected" (Kano *et al.* 1993a).

Elements	Allende	Nuevo Mercurio	Canyon Diablo
Al (%)	16.0 ± 0.1	2.18 ± 0.02	0.0762 ± 0.0034
Ca (%)	0.49 ± 0.23	nd	nd
Mg (%)	12.5 ± 1.6	5.73 ± 1.00	nd
Mn (ppm)	81.4 ± 52.8	452 ± 30	nd
Ti (ppm)	770 ± 679	566 ± 99	nd
V (ppm)	230 ± 9	285 ± 9	3.76 ± 1.81
Fe (%)	1.89 ± 0.02	4.32 ± 0.02	1.05 ± 0.02
Ni (ppm)	145 ± 36	<12.8	434 ± 24
Co (ppm)	21.5 ± 0.3	5.68 ± 0.19	60.9 ± 0.4
Cr (ppm)	390 ± 7	11200 ± 100	15.4 ± 1.2
Au (ppm)	1.87 ± 0.02	0.0097 ± 0.0013	0.956 ± 0.006
Sb (ppb)	586 ± 28	13.7 ± 2.0	121 ± 6
W (ppb)	696 ± 172	nd	1910 ± 470
Re (ppb)	580 ± 74	nd	230 ± 19
Os (ppm)	4.15 ± 0.89	nd	1.66 ± 0.37
Ir (ppm)	5.46 ± 0.06	0.15 ± 0.005	1.10 ± 0.01
Mo (ppb)	<643	nd	1920 ± 1100
Ru (ppm)	1.95 ± 0.84	nd	1.79 ± 0.72
Pt (ppm)	1.78 ± 1.47	nd	0.792 ± 0.64
La (ppb)	225 ± 38	114 ± 20	214 ± 26
Sm (ppb)	113 ± 9	nd	13.8 ± 2.2
Eu (ppb)	84.8 ± 34	nd	nd
Yb (ppb)	nd	nd	66 ± 43
Sc (ppm)	77.8 ± 0.3	3.38 ± 0.03	37.4 ± 11..4

In isotopic analyses of Ru-isotopes, no remarkable anomalies were detected now by a highly sensitive, thermal ionization mass spectrometer.* Generally, acid residues are considered to be presolar grains and/or secondary inflown fractions of diverse nucleosynthetic origins arising from an interstellar medium during the condensation periods of the solar system. They are also thought to be components of the primary condensates from the cooling gas of the solar composition. The diverse compositions of acid residues between various meteorites might arise from differences in proportion and composition of these constituents arising due to different environmental conditions in space. Thus, the acid residues of various meteorites are considered to contain some components that depend strongly on the bulk composition of each meteorite or the parent bodies of these meteorites.

G. Studies on Magnetic Fractions Bearing High ($^3He/^4He$) Values

As described in Section 5.3, fairly large ($^3He/^4He$) values have been found for deep sea sediments (Merrihue 1964; Amari *et al.* 1985; Fukumoto *et al.* 1986; Takayanagi *et al.* 1987). This could be the result of the small, meteoroidal grains having been subjected to solar ^3He winds and flare particles while in the vicinity of Sun before their eventual decent to the Earth and their becoming embedded into what are now deep sea sediments. It is remarkable that, given the small size of the grains, such implanted ions were not ablated and did not evaporate away during periods of the Earth's atmospheric heating. In their fractions, the most likely carrier of the solar ^3He ions was the beta-meteoroid, which was blown out of the vicinity of Sun after sublimation by sun light heatings (see Section 2.2). The terrestrial ($^3He/^4He$) values for MORB and/or the upper mantle layers are one-tenth or less the sedimental values. Systematic studies are presently being carried out to prove this hypothesis.

A large volume of deep sea sediments, which were taken at depth more than 4700 meters off the Hawaiian Islands in 1979 by R/V Hakurei Maru, was separated using sieves having pores 20 μm and 74 μm of the openings. These sizes correspond with the critical sizes of iron and of stony meteoroids which had not been heated to 300 K during the periods of atmospheric heating, respectively (Hunter and Parkin 1960). The magnetic fractions were gathered using a rod-magnet, whose electric currents were set at DC 2A, 6A and 10A. We have nine portions separated by sizes ($S < 20$ μm, 20 μm $< S <$ 74 μm, $S >$ 74 μm) and by electric currents of 2A, 6A and 10A. Almost all stony spherules contain enough iron that they could be gathered by magnets. Thereafter, their chemical

*Ru stable isotopic compositions are given here; Ru-96 = 5.52% (p-process), Ru-98 = 1.88% (p-process), Ru-99 = 12.7% (r and s processes), Ru-100 = 12.6% (s-process only), Ru-101 = 17.0% (r and s processes), Ru-102 = 31.6% (r and s processes) and Ru-104 = 18.7% (r-process only).

Table A-5. Chemical compositions of the magnetic fractions sorted by sizes and amplitude of electric current (Yamakoshi *et al.* 1992).

Elements	F-2A	F-6A	F-10A	M-2A	M-6A	M-10A	C-2A	C-6A	C-10A
Fe (%)	37.3 ± 1.6	31.7 ± 1.8	25.0 ± 1.7	35.3 ± 1.5	23.0 ± 1.4	14.9 ± 1.7	32.3 ± 1.3	12.5 ± 1.5	8.09 ± 1.4
Ni (%)	0.58 ± 0.17	0.34 ± 0.16	—	0.43 ± 0.16	0.11 ± 0.03	—	0.28 ± 0.15	—	—
Co (ppm)	185 ± 12	136 ± 11	125 ± 22	156 ± 08	155 ± 13	43.9 ± 20	178 ± 20	56.0 ± 22	74.0 ± 27
Cr (ppm)	1160 ± 40	1400 ± 50	979 ± 50	812 ± 32	1130 ± 12	604 ± 80	923 ± 47	365 ± 75	60.0 ± 60
Au (ppb)	24.5 ± 7.1	7.0 ± 4.7	—	13.0 ± 4.5	—	—	—	—	—
Ga (ppm)	28 ± 7	38 ± 11	22 ± 6	23 ± 6	28 ± 8	26 ± 8	30 ± 8	26 ± 8	22 ± 7
W (ppm)	2.4 ± 0.5	1.1 ± 0.2	2.6 ± 0.7	1.7 ± 0.3	2.2 ± 0.4	1.5 ± 0.4	1.8 ± 0.4	1.7 ± 0.6	4.1 ± 0.7
Os (ppb)	711 ± 387	—	—	474 ± 350	—	—	1110 ± 580	—	—
Ru (ppb)	1500 ± 950	669 ± 482	—	1130 ± 660	912 ± 803	—	1390 ± 900	—	—
Ir (ppb)	50.3 ± 10.7	—	—	—	—	—	—	—	—
Pt (ppm)	2.5 ± 2.4	—	—	—	—	—	—	—	—
La (ppm)	12.0 ± 5.5	24.0 ± 10.9	29.5 ± 13.5	17.0 ± 7.8	54.4 ± 24.8	62.0 ± 28.2	21.6 ± 9.9	69.5 ± 31.6	72.6 ± 33.0
Sm (ppm)	4.2 ± 0.24	7.3 ± 0.47	10.5 ± 0.6	6.7 ± 0.4	18.0 ± 1.1	18.7 ± 1.2	7.83 ± 0.5	22.0 ± 1.4	23.5 ± 1.5
Eu (ppm)	0.81 ± 0.05	1.9 ± 1.3	2.2 ± 0.2	1.6 ± 0.1	4.4 ± 3.0	4.1 ± 2.8	1.8 ± 0.1	5.1 ± 3.4	4.7 ± 3.2
Yb (ppb)	632 ± 448	748 ± 422	807 ± 571	648 ± 449	1750 ± 960	2660 ± 1470	495 ± 361	1630 ± 930	1330 ± 750
Sc (ppm)	15.5 ± 1.6	21.1 ± 2.6	20 ± 2.3.1	15.2 ± 1.6	21.4 ± 2.8	27.6 ± 3.7	17.8 ± 1.8	24.0 ± 3.3	28.5 ± 3.8

Table A-5. (Continued).

[pneumatic tube irradiation]									
Al (%)	2.73 ± 0.02	3.46 ± 0.03	3.95 ± 0.03	3.06 ± 0.02	4.01 ± 0.04	4.88 ± 0.04	3.45 ± 0.02	4.84 ± 0.04	5.89 ± 0.04
Mg (%)	5.26 ± 0.99	4.35 ± 0.87	3.94 ± 0.63	4.28 ± 0.71	5.99 ± 1.13	2.35 ± 0.65	3.94 ± 0.74	5.11 ± 1.14	4.22 ± 0.99
Mn (%)	0.40 ± 0.03	0.40 ± 0.03	0.35 ± 0.02	0.35 ± 0.02	0.26 ± 0.02	0.25 ± 0.20	0.28 ± 0.02	0.27 ± 0.02	0.23 ± 0.02
Ti (%)	5.20 ± 0.67	5.11 ± 0.66	4.70 ± 0.61	4.87 ± 0.63	3.48 ± 0.47	2.89 ± 0.39	4.46 ± 0.58	3.18 ± 0.43	1.58 ± 0.22
V (ppm)	1480 ± 40	1380 ± 40	1230 ± 40	1300 ± 40	851 ± 28	661 ± 22	1200 ± 40	667 ± 23	334 ± 13
Ca (%)	0.55 ± 0.14	0.80 ± 0.17	1.11 ± 0.20	0.37 ± 0.10	1.04 ± 0.26	2.22 ± 0.39	1.35 ± 0.39	1.81 ± 0.42	2.04 ± 0.37

More extensive laboratory simulation experiments of heavy ion implantation into iron and stony minerals must be done in the future (Futagami *et al.* 1990).

compositions and ($^{3}He/^{4}He$) ratios were determined. In this work, the pelagic sediments were first diluted and homogenized with a large volume of pure distilled water using an electric stirrer and then the muddy water was separated gradually, using sieves, into three portions; 20 μm > S (fine grained, F), 74 μm > S > 20 μm (medium grained, M) and S > 74 μm (coarse grained, C). The magnetic fractions were then gathered using rod-magnets whose electric currents were DC 2A (named as 2A), 6A (as 6A) and 10A (as 10A), respectively. A pure-iron rod, having a diameter of 15 mm and a length of 200 mm, was coiled with a covered line, 800 turns, so that the coil's resistance was ca 1.5 ohms for a DC voltage. During stirring of the muddy water the temperature of the rod magnets increased because of heat coming from the coil, so that current values decreased by ~10%. Thus, the supplied voltage to an AC-DC convertor was adjusted repeatedly for to compensate for this effect.

It has been preliminarily proved, by noble gas spectrometry, that sedimental samples do not lose their own gas-fractions when they are washed by ultra-sonic vibrators. Thus, separated fractions were washed in a large volume of pure alcohol using an ultra-sonic cleaner and the clustered debris and aggregates were broken and separated into fines. The sieving procedures were repeated.

Generally speaking, in the portions separated by rod-magnets the 2A fractions contained brilliant black, iron spherules. Magnetite grains of irregular forms were also found, except in F-2A. The 6A fractions also contained iron and stony spherules, and in the 10A fractions many micro-ferromanganese nodules and colorful rocky fines were apparent. In the F-2A fraction various types of angular and/or aggregates grains were present. Spherical grains could scarcely be seen. After all the procedures had been performed, the individual fractions were treated by a de-magnetizer for a few minutes.

In this work, two experimental procedures were performed: non-destructive neutron activation analyses [INAA], for chemical composition, and mass-spectrometry for ($^{3}He/^{4}He$) determination. INAA was carried out at a reactor, TRIGA II, whose neutron flux was 3×10^{12} (n/sec.cm^{2}). Micro-fine (<20 μm) powdered Allende was used as a reference sample. The respective samples were measured by a Ge (Int) detector system and peak area and decay analyses were performed with a personal computer. Helium isotopes were analyzed following the conventional technique of noble gas mass spectrometry, which is described in detail elsewhere (Takaoka 1976). In Tables A-5 and A-6, the INAA results (Kano *et al.* 1993b) and He-gas analysis (Yamakoshi *et al.* 1991) are shown, respectively.

In this work, the 2A-fractions contained abundant noble metals and also had associated high ($^{3}He/^{4}He$) values. These ($^{3}He/^{4}He$) values were obtained from the mixtures of extaterrestrial dust saturated with solar ^{3}He as much as 1–10 ppm. These results are very similar to those obtained by Yamakoshi *et al.* (1981).

Table A-6. He gas mass spectrometry of the magnetic fractions gathered from pelagic sediments (Takaoka 1986; Yamakoshi *et al.* 1992).

Sample code (mgm)	($^3He/^4He$) $\times 10^{-4}$	^4He $\times 10^{-7}$ (ml)	^4He $\times 10^{-5}$ (ml STP/gm)	^3He $\times 10^{-9}$ (ml STP/gm)
F-2A (12.468)	2.5 ± 0.2	2.4 ± 0.3	1.9 ± 0.7	4.8 ± 0.2
M-2A (34.776)	2.2 ± 0.1	9.9 ± 1.4	2.8 ± 1.0	6.1 ± 2.3
M-6A (20.398)	1.4 ± 0.3	3.3 ± 0.5	1.6 ± 0.6	2.3 ± 0.9
M-10A (29.163)	1.7 ± 0.1	3.4 ± 0.5	1.2 ± 0.4	1.9 ± 0.7
C-2A (34.739)	1.9 ± 0.1	5.5 ± 0.8	1.6 ± 0.6	3.0 ± 1.1

Fe, Ni, Co, Mn, Ti and V were found to be most abundant in the 2A samples, but as sizes and amplitude of the current increased, the relative abundance of these components decreased. Os, Ru and Au were found only in the 2A fractions. Ir and Pt were found in F-2A alone. The content of REE (La, Sm and Yb) as well as that of Sc, Al and Ca contents decreased as sizes and currents increased. Other fractions (W, Mg, Eu, Ga, Cr) did not exhibit the same correlations with size and current.

H. Brownlee's Particles Found in Deep Sea Sediments

In July 1993, three Brownlee's particles were discovered in the magnetic fractions gathered from deep sea sediments, dredged by R/V Hakurei-Maru off the Hawaiian Islands at a depth of more than 4500 meters. Suzuki *et al.* (1993), have been searching for extraterrestrial grains, using a highly sensitive X-ray micro-analyzer among several hundred grains whose sizes exceeded 74 μm of a magnetic fraction named C-10A (see Appendix G). The chemical composition of a typical large grain, having a size of 100 × 95 μm, is shown in Fig. A-9.

The data are averaged using the results of 10 point measurements over all sample grains. It may be noted that the nickel analytical spots are not concentrated in special regions, but distributed widely over the sample surface. Cu and Zn peaks originate from the base-material (brass). The preliminary data shown in Fig. A-9 are tabulated here (Suzuki *et al.* 1993): [Si] = 38.3, [Fe] = 32.0, [Mg] = 17.2, [Ca] = 2.7, [Cr] = 2.0, [Ti] = 0.72, [Ni] = 0.58 and [Al] = 2.4 (wt %).

In Fig. A-10, typical data of Brownlee's (stratospheric) particles, obtained and compiled by the CDPET of the Johnson Space Center of NASA (see Section 4), is shown, whose code is L2005 G.

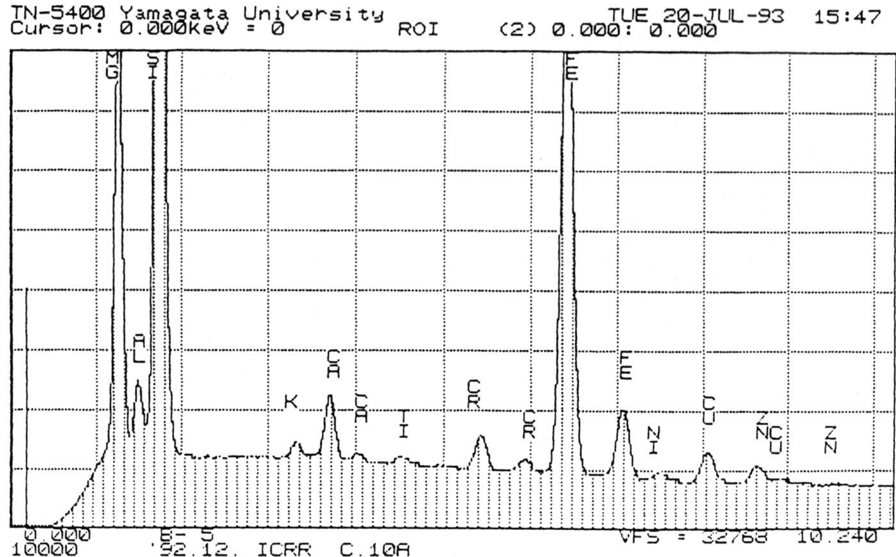

Fig. A-9. A typical composition of a grain obtained from C-10A (Suzuki *et al.* 1993).

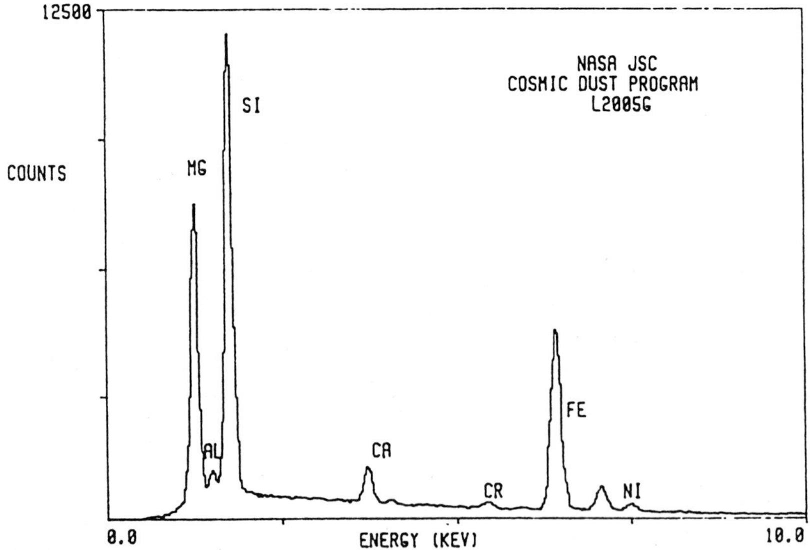

Fig. A-10. A typical data diagram of the stratospheric dust, which is assigned as cosmic origin (Johnson Space Center, NASA).

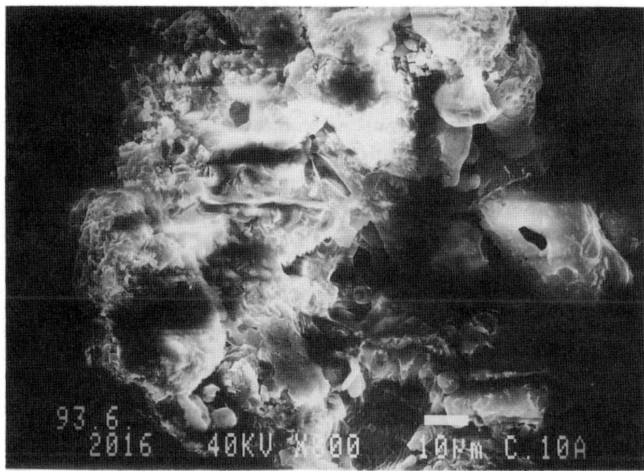

Fig. A-11. A picture of the grain whose composition is shown in Fig. A-9.

Both diagrams had surprising similarity to one another, however, the sample used by NASA is quite small (~5 μm). In Fig. A-11, a picture obtained using a transmission electromicroscope [JEM-2000FX] is shown.

Because of the discovery of these fragile and delicate dust particles in the deep sea magnetic fractions, it can be expected that future research will focus on the parent-child relationship between meteorites and interplanetary dusts with respect to the precise determinations of their chemical and isotopical compositions.

REFERENCES

Amari S. and Ojima M. 1985 *Nature* **317** 520.

Esat T. M., Brownlee D. E., Papanastassiou D. A. and Wasserburg G. J. 1979 *Science* **206** 190.

Fukumoto H., Nagao K. and Matsuda J. 1985 *Geochim. Cosmochim. Acta* **50** 2245.

Futagami T., Ojima M. and Nakamura Y. 1990 *Earth Planet. Sci. Lett.* **101** 63.

Gruen E. T., Baguhl M., Fechtig H., Hanner M. S., Kissel J., Lindblad B. A., Linkert G., McDonnell J. A. M., Morfill G. S., Siddique N. and Zook H. A. 1992 Proc. HIS Workshop (Katlenburg-Lindau).

Gruen E. T., Zook H. A., Baguhl M., Baloch A., Bame S. J., Fechtig H., Forsyth R., Hanner M. S., Horanyi M., Kissel J., Lindblad B. A., Linkert D., Linkert G., Mann I., McDonnell J. A. M., Morfill G. E., Phillips J. L., Polanskey C., Schwehm G., Siddique N., Staubach P., Svestka J. and Taylor A. 1993 *Nature* **362** 428.

Hunter W. and Parkin D. W. 1960 *Proc. Roy. Soc. A* **255** 382.

Kano N., Yamakoshi K., Matsuzaki H. and Nogami K. 1993a *Proc. Antarct. Meteorit.* 16.

Kano N., Matsuzaki H., Yamakoshi K., Takaoka N. and Nogami K. 1993b unpublished.

Keller H. U., Richter K. 1991 "*Origin and Evolution of Interplanetary Dust*" eds. A. C. Levasseur-Regourd and H. Hasegawa (Kluwer Academic Publishers, Tokyo) 229.

Kumar P. and Goel P. S. 1991 *Geochem. J.* **25** 399.

Li Chun-lai and Ouyang Ziyuan 1992 Abstracts Sympos. Antarctic Meteorites. 36.

Merrihue C. 1964 *Ann. New York Acad. Sci.* **119** 351.
Misawa K. and Tsuchiyama A. 1990 Abstracts Sympos. Antarctic Meteorites **15** 170.
Misawa K., Yamakoshi K. and Nakamura N. 1992 *Geochem. J.* **26** 29.
Misawa K., Yamakoshi K., Nogami K., Nakamura N. and Yamamoto K. 1993 ibid. **26** 197.
Ping Kong and Chifang Chai 1992 Abstracts Sympos. Antarctic Meteorites 35.
See T., Albrooks M., Simon C. and Zolensky M. 1990 A Preliminary Report of LDEF, "*Meteoroids and Debris Impact Features Documented on the LDEF*", JSC #24608, NASA.
Suzuki Y., Noma M., Sakurai H., Yamakoshi K., Matsuzaki H., Kano N. and Nogami K. 1993 unpublished.
Sykes M. V., Lebofsky L. A., Hunter D. M. and Low F. J. 1986 *Science* **232** 1115.
Takaoka N. 1976 *Mass spectrometry* **24** 73.
Takayanagi M. and Ojima M. 1987 *J. Geophys. Res.* **92(B12)** 12531.
Yamakoshi K. 1992 unpublished.
Yamakoshi K., Nogami K. and Shimamura T. 1981 ibid **B4** 3129.
Yamakoshi K., Ohashi H., Noma M., Sakurai H., Nakashima K., Nogami K. and Omori R. 1993 Abstracts 24th Annual Lunar Planet. Science Conf. (Houston).
Zinner E., Tang M. and Anders E. 1987 *Nature* **330** 730.
Zinner E., Amari S., Anders E. and Lewis R. S. 1991a ibid. **349** 51.
Zinner E., Amari S. and Lewis R. S. 1991b Abstracts LPSC XXII 1553.

Subject Index